The
Reference Shelf®

Dinosaurs

Edited by

John J. Meier

The Reference Shelf
Volume 83 • Number 2
The H.W. Wilson Company
New York • Dublin
2011

The Reference Shelf

The books in this series contain reprints of articles, excerpts from books, addresses on current issues, and studies of social trends in the United States and other countries. There are six separately bound numbers in each volume, all of which are usually published in the same calendar year. Numbers one through five are each devoted to a single subject, providing background information and discussion from various points of view and concluding with a subject index and comprehensive bibliography that lists books, pamphlets, and abstracts of additional articles on the subject. The final number of each volume is a collection of recent speeches, and it contains a cumulative speaker index. Books in the series may be purchased individually or on subscription.

Library of Congress has cataloged this serial title as follows:

Dinosaurs / edited by John J. Meier.
    p. cm. -- (The reference shelf ; v. 83, no. 2)
  Includes bibliographical references and index.
  ISBN 978-0-8242-1107-3 (alk. paper)
  1. Dinosaurs. 2. Paleontologists. 3. Dinosaurs--Evolution. 4. Dinosaurs--Extinction. I. Meier, John J.
  QE861.4.D562 2011
  567.9--dc22
                    2011007540

Cover: A Tyrannosaurus Rex dwarfs a small girl at the media call for "Walking With Dinosaurs-The Arena Spectacular" at The Entertainment Quarter on October 18, 2010 in Sydney, Australia. The arena show featuring life-sized dinosaurs returns to Australia after a successful world tour. (Photo by Brendon Thorne/Getty Images)

Visit H.W. Wilson's Web site: www.hwwilson.com

Printed in the United States of America

Contents

Preface

When we're children, no scientific topic captures our imagination quite like dinosaurs. Perhaps it's because so much imagination is required to dream up sprawling, prehistoric landscapes and populate them with ravenous raptors and swooping pterodactyls. While most of us eventually outgrow our obsession with *Jurassic Park*, some ride dinosaurs, figuratively speaking, into adulthood and become paleontologists. But professional bone hunters aren't the only ones concerned with these prehistoric beasts. As this book makes clear, amateur fossil hunters, biologists, journalists, physicists, and students also make important contributions to the field. And yet, despite all the discoveries these and other people have made, certain intriguing mysteries remain. What killed the dinosaurs? Are they the ancestors of birds? Was *Tyrannosaurus rex* a ferocious predator or glorified chicken?

The articles collected in this volume of The Reference Shelf are meant to give an overview of what we've come to learn about dinosaurs. Because our understanding grows and changes with each new discovery, the selections were culled exclusively from the past decade. Most are examples of what some would call "popular" science writing: pieces intended for anyone interested in the topic, not just experts.

Entries in the first chapter, "What Were the Dinosaurs?" offer an introduction to dinosaurs. The section opens with an information-rich encyclopedia entry, then moves to lighter, flashier pieces that consider dinosaurs' place in popular culture. The final selections discuss the state of modern paleontology, a discipline that has some scientists putting down their shovels and pickaxes and moving into the laboratory.

Articles in the second chapter, "Bone Sharps, Fossil Hunters, and Dinosaur Experts," center on those who "chase" dinosaurs for a living. These selections take us on a world tour, jumping from the badlands of the western United States to Australia, focusing on day-to-day life in the field, and introducing the scientists and private prospectors who are among today's fossil seekers. The articles touch on some of the motivations—some scientific, others financial—for going dinosaur hunting.

Dinosaurs were once depicted as slow and hulking, like modern lizards, but over the past few decades, that view has changed. Many researchers now insist they were quick, feathered creatures—precursors to modern birds. Pieces in the third chapter, "Of a Feather? The Bird-Dinosaur Link," explore these theories and the evidence supporting them. These articles reference everything from traditional fossil and imprint research to emerging techniques that enable biologists to reconstruct dinosaur proteins.

There is still a great deal of controversy regarding dinosaur extinction, and articles in the fourth chapter, "What Killed the Dinosaurs?" offer numerous explanations. Many mass extinctions and catastrophes occurred roughly 65 million years ago, but scientists have yet to agree on which led to the demise of the dinosaurs. As the selections explain, scientists have used geological formations and fossil records to identify events and ecological changes that might be to blame. If dinosaurs were done in by a single asteroid, as many believe, there remains the question of where on Earth it struck, as numerous craters have been found.

In such a small collection of articles, it is difficult to capture the true breadth and frequency of advances being made in the field of dinosaur research. Nevertheless, entries in the fifth and final chapter, "New Discoveries," detail some of the more interesting findings to have emerged in recent years. The first article is not actually about dinosaurs, but rather ancient marine reptiles, and it seeks to explain the differences between the two. The subsequent selections highlight research supporting the bird-dinosaur link, as well as new types of non-fossil evidence, such as footprints and soft tissue.

In closing, I would like to thank Raymond Barber and Joseph Miller for bringing me in as a science book nominator for H.W. Wilson, as well as trusting me with additional opportunities, such as editing this book. I must also thank Paul McCaffrey for being a very communicative editor on this project, his H.W. Wilson colleagues Kenneth Partridge and Richard Stein for all their help, and my wife, Mary, for being a sometime copyeditor and amazing life partner. Thanks, too, to my son, Henry, for loving dinosaurs, even as a two-year-old. (Pachy-cefalo-saurus!) Finally, I want to dedicate this volume to my grandfather, Jack, who left this world at the end of 2010. He taught me the value of both hard work and laughter.

John J. Meier
April 2011

1

What Were the Dinosaurs?

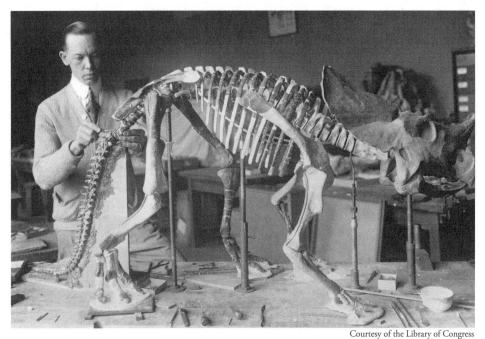

Courtesy of the Library of Congress

Norman Ross of the division of Paleontology, National Museum, preparing for exhibition the skeleton of a baby dinosaur some seven or eight million years old. LC-USZ62-127774

A mounted *Stegosaurus* specimen at the Smithsononian National Museum of Natural History in Washington, D.C.

Editor's Introduction

Mankind's study of dinosaurs stretches back more than 150 years, so there is a rich history and a great deal of knowledge to cover. Because many definitions and terms associated with the field have stood the test of time, this introductory chapter focuses less on tracing the history of the science than providing a broad overview of the subject.

The first selection is the encyclopedia entry "Dinosauria" from *AccessScience*, the on-line version of the *McGraw-Hill Encyclopedia of Science & Technology*. Written by Kevin Padian and Paul M. Barrett, this article is longer and slightly more detailed than the average dictionary or encyclopedia entry. After briefly summarizing the history of dinosaur study, Padian and Barrett delve into the important subjects of classification and taxonomy, describing some of the groups of dinosaurs that will surface again in later chapters.

In the next piece, "Dinosaur Dreams: Reading the Bones of America's Psychic Mascot," *Harper's* contributing editor Jack Hitt weaves scientific history with popular culture as he reviews the most significant events in the history of dinosaur study. Telling the story from 1677, when the first-ever dinosaur bone—labeled "Scrotum humanum" by artist Richard Brookes—was identified, up to the present day, Hitt uses great wit and insight to blow dust from the pages of history. Calling dinosaurs "distinctly American," Hitt discusses how world events and cultural trends have shaped the public's view of these prehistoric creatures.

In the subsequent article, "Extreme Dinosaurs," the renowned American author John Updike explores some of the odder dinosaur specimens, pondering the evolutionary benefits of "huge triple claws," "horizontally protruding front teeth," and other anatomical curiosities observed in the fossil record. Updike's musings highlight just how diverse dinosaurs were, and writing in his usual erudite style, he adds color to his scientific explanations.

In "Flesh & Bone: A New Generation of Scientists Brings Dinosaurs Back to Life," former *National Geographic* science columnist Joel Achenbach introduces the "new" sciences related to dinosaur study. From measuring the force of an alligator's jaws to modeling dinosaur joint movement on the computer, there are a multitude of ways scientists are using technology to enhance their understanding of dinosaurs. Modern paleontologists can spend as much time in the laboratory as in the field, and Achenbach discusses both the questions they study and the tools they use to answer them.

Erik Stokstad, a staff writer and editor for the prestigious journal *Science*, ends the chapter with "Dinosaurs Under the Knife," a look at the new enterprise of studying dinosaur fossil cells and tissue. Though fossilized bone tissue is difficult to analyze, researchers are able to look for patterns and differences between separate specimens. They can also make hypotheses about how dinosaurs behaved based on the growth and stress found in their bones.

Dinosauria[*]

By Kevin Padian and Paul M. Barrett
AccessScience from McGraw-Hill

The term, meaning awesome reptiles, that was coined by the British comparative anatomist R. Owen in 1842 to represent three partly known, impressively large fossil reptiles from the English countryside: the great carnivore *Megalosaurus*, the plant-eating *Iguanodon*, and the armored *Hylaeosaurus*. They were distinct, Owen said, not only because they were so large but because they were terrestrial (unlike mosasaurs, plesiosaurs, and ichthyosaurs); they had five vertebrae in their hips (instead of two or three like other reptiles); and their hips and hindlimbs were like those of large mammals, structured so that they had to stand upright (they could not sprawl like living reptiles). Owen's diagnosis was strong enough to be generally valid, with some modification, 150 years later. His intention in erecting Dinosauria, though, as A. J. Desmond noted, seems to have been more than just the recognition of a new group: By showing that certain extinct reptiles were more "advanced" in structure (that is, more similar to mammals and birds) than living reptiles, he was able to discredit contemporary evolutionary ideas of the transmutation of species through time into ever-more-advanced forms (progressivism).

Over the next several decades, dinosaurs were discovered in many other countries of Europe, for example, *Iguanodon* in Belgium and *Plateosaurus* in Germany, but rarely in great abundance. The first dinosaur discoveries in the United States were from New Jersey as early as the 1850s [*Trachodon*, 1856; *Hadrosaurus*, 1858; *Laelaps* (=*Dryptosaurus*), 1868]. The genus *Troodon*, based on a tooth from Montana, was also described in 1856, but at the time it was thought to be a lizard. Spectacular discoveries of dinosaurs from the western United States and Canada began in a great rush in the 1870s and 1880s and continued into the early 1900s. From 1911 to 1914, expeditions into German East Africa discovered some of the largest dinosaurs ever collected: *Brachiosaurus* and *Tornieria*, among many others. In the 1920s, expeditions into the Gobi desert of Mongolia brought back new and unusual dinosaurs such as *Protoceratops*, and the first eggs immediately recognizable as dinosaurian. For various reasons, such grand expeditions dwindled until the late 1960s, when renewed activities in the western United States and Canada, Argentina, southern Africa, India, Australia, China, Mongolia, and even Antarctica uncovered dozens of new dinosaurs. The rate of new discoveries reached an estimate of one new find every 6 weeks (intriguing new ones include *Microraptor*, *Mononykus*, *Eoraptor*, and *Brachytrachelopan*) during the mid-1990s, a pace of discovery that has continued (or possibly been exceeded) in the early twenty-first century. The finds since the 1960s may represent up to 50% of all dinosaurs currently known to have existed.

HIP STRUCTURE

As dinosaurs became better known, their taxonomy and classification developed, as well as their diversity. In 1887, H. G. Seeley recognized two quite different hip structures in dinosaurs and grouped them accordingly. Saurischia, including the carnivorous Theropoda and the mainly herbivorous, long-necked Sauropodomorpha, retained the generalized reptilian hip structure in which the pubis points down and forward and the ischium points down and backward. The remaining dinosaurs have a pubis that has rotated to point down and backward, thus extending parallel to the ischium; this reminded Seeley of the configuration in birds, and so he named this group Ornithischia. However, the ornithischian pubis is only superficially similar to that of birds, which are descended from, and are thus formally grouped within, Saurischia. Seeley's discovery, in fact, only recognized the distinctiveness of Ornithischia, and he concluded that Saurischia and Ornithischia were not particularly closely related. Even within Saurischia, there were general doubts that Sauropoda and Theropoda had any close relationship; eventually, the word "dinosaur" was used mainly informally by paleontologists.

However, this situation has been reversed. In 1974, Robert Bakker and Peter Galton argued that there were a great many unique features, including warm-bloodedness, which characterized the dinosaurs as a natural group (also including their descendants, the birds). Although this scheme was debatable in some particulars, it spurred renewed studies anchored in the new methodology of cladistic analysis of phylogenies (evolutionary relationships). A 1986 analysis listed nine uniquely derived features (synapomorphies) of the skull, shoulder, hand, hip, and hindlimb that unite Dinosauria as a natural group; this analysis has been since modified and improved, and today Dinosauria is universally accepted as a natural group, divided into the two monophyletic clades, Ornithischia and Saurischia.

EARLIEST DINOSAURS

Dinosaurs are archosaurs, a group that comprises living crocodiles, birds, and all of the living and extinct descendants of their most recent common ancestor (**Fig.**

1). The closest relatives of dinosaurs, which evolved with them in the Middle and Late Triassic (about 240–225 million years ago), include the flying pterosaurs and agile, rabbit-sized forms such as *Lagosuchus* and *Lagerpeton*. The common ancestor of all these forms was small, lightly built, bipedal, and probably an active carnivore or omnivore. Somewhat larger, with skulls ranging 15–30 cm (6–12 in.) in length, were *Eoraptor* and *Herrerasaurus* from the Late Triassic of Argentina and *Staurikosaurus* from the early Late Triassic of Brazil (**Fig. 2**). When the latter two genera were first described in the 1960s, they were thought to be primitive saurischian dinosaurs. The relationships of these animals are contentious, however, with some authors placing them within Theropoda, others positioning them at the base of Saurischia, and a further group of specialists suggesting that they were outside the group formed by Saurischia plus Ornithischia. These Late Triassic genera testify to a burst of evolutionary change at this interesting time in vertebrate history, and show that a variety of taxa very close to the origin of dinosaurs appeared during this interval. The first definite ornithischians and saurischians appear at almost the same time, though dinosaurs remained generally rare and not very diverse components of terrestrial faunas until the beginning of the Jurassic Period (about 200 million years ago).

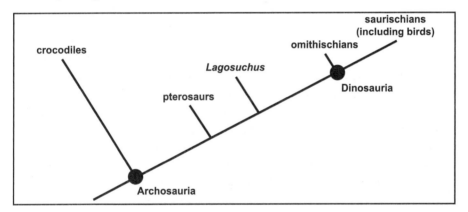

Fig. 1 Simplified phylogenetic tree of the archosaurian reptiles.

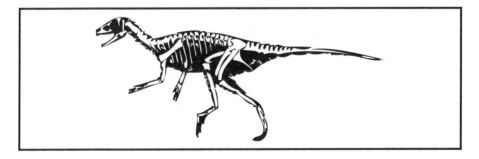

Fig. 2 *Eoraptor*, an early dinosaur or dinosaurian relative from the Late Triassic of Argentina.

An area of great interest is how the dinosaurs and their closest relatives differ from their contemporaries. Their posture and gait hold some important clues. Like *Lagosuchus, Lagerpeton,* and their other close relatives, the first dinosaurs stood upright on their back legs. The head of the thigh bone (femur) angled sharply inward to the hip socket, which was slightly open (not backed by a solid sheet of bone as in other reptiles). The femur moved like a bird's, in a nearly horizontal plane; the shin bone (tibia) swung back and forth in a wide arc, and the fibula (the normally straplike bone alongside the tibia) was reduced because the lower leg did not rotate about the knee, as a crocodile's or lizard's does. The ankle, too, had limited mobility: it formed a hinge joint connecting the leg to long metatarsals (sole bones), which were raised off the ground (in a stance called digitigrady, or "toe-walking"). All of these features can be seen today in birds, the living descendants of Mesozoic dinosaurs. Because the first dinosaurs were bipedal, their hands were free for grasping prey and other items, and the long fingers bore sharp, curved claws. The neck was long and S-shaped, the eyes large, and the bones lightly built and relatively thin-walled.

ORNITHISCHIA

Ornithischians (**Fig. 3**) are a well-defined group characterized by several unique evolutionary features; the entire group was analyzed cladistically in 1986, and the resulting phylogeny has been the basis of all later work. Ornithischians have a predentary bone, a toothless, beaklike addition to the front of the lower jaw that, like the front of the upper jaw, probably had a horny covering in life (**Fig. 4**). This appears to have been an adaptation for plant eating. The jaw joint was set below the occlusal plane, a nutcracker-like arrangement interpreted as increasing leverage for crushing plant material. The teeth were set in from the side of the jaw, suggesting the presence of fleshy cheeks to help sustain chewing. The cheek teeth were broad, closely set, and leaf-shaped, and were often ground down to a shearing surface. In the hip, the pubis pointed backward. In all but the most generalized ornithischians, a new prong on the pubis was developed from the hip socket, upward, forward, and outward. This may have provided a framework to support the guts or to anchor the hindlimb muscles.

The most generalized ornithischians known are the fragmentary *Pisanosaurus* from the Late Triassic of Argentina and the small *Lesothosaurus* from the Early Jurassic of South Africa. In the major ornithischian radiation, Thyreophora branch off first, and Cerapoda are divided into Ornithopoda and Marginocephalia. All known ornithischians are either omnivores or herbivores.

Fig. 3 Relationships of the ornithischian dinosaurs. The base of this diagram is linked to Fig. 12. (*Modified from P. C. Sereno, The evolution of dinosaurs, Science, 284:2137–2147, 1999*)

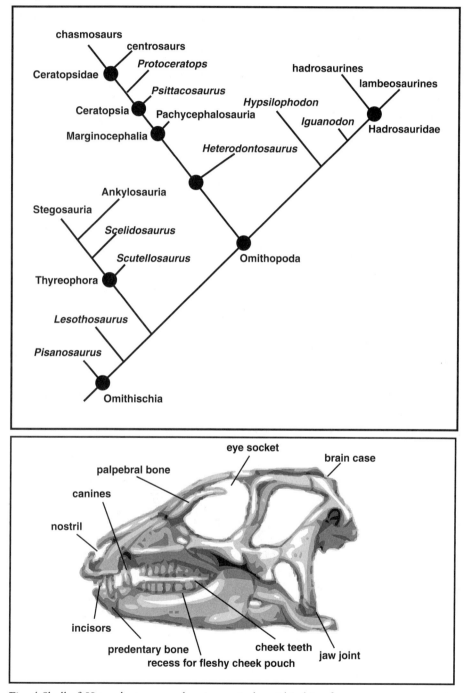

Fig. 4 Skull of *Heterodontosaurus*, showing typical ornithischian features.

THYREOPHORA

Stegosaurs and ankylosaurs were distinguished by their elaborate armor of modified dermal bone. The earliest thyreophorans, such as the small, bipedal *Scutellosaurus* and the slightly larger, probably quadrupedal *Scelidosaurus* (both from the Early Jurassic), had such scutes all over their bodies, but not as elaborate or distinctive as those of the larger, fully quadrupedal stegosaurs and ankylosaurs.

Stegosaurs

Stegosaurs first appear in the Middle Jurassic of China and Europe, and were undoubtedly widespread by then, though the group never seems to have been particularly diverse: 14 genera are known, mostly from the Late Jurassic and Early Cretaceous. They had the reduced ancestral armor pattern of Thyreophora, losing the scutes on the sides of the body and retaining only a row along either side of the vertebral column (**Fig. 5**). Originally this paravertebral armor was spiky or conelike (*Huayangosaurus*), but in some forms (*Kentrosaurus*) the armor forward of the middorsal region became platelike polygons, and in *Stegosaurus* the now much-enlarged, subtriangular plates occupied all but the last two terminal tail positions, which were still spiky. All well-known stegosaurs (except *Stegosaurus* itself) seem to have had a long spine projecting upward and backward from the shoulder as well. *Stegosaurus* also had a complement of bony ossicles in its throat region. In all stegosaurs the heads were relatively small and the teeth few and diamond-shaped. Stegosaurs had small brains, ranging from a walnut to a billiard ball in size. An expansion of the neural canal in the hip region was once interpreted as a "second brain" to control the hindlimbs and tail, but it is only a cavity to accommodate the sacral nerve plexus and a large glycogen ("fat") storage organ like those of birds and some other reptiles.

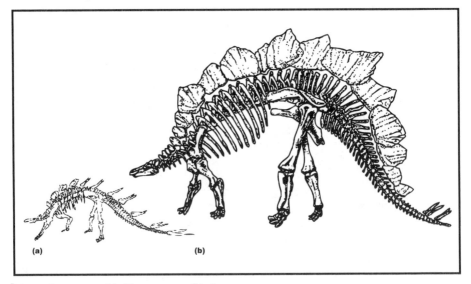

Fig. 5 Stegosaurs. (a) *Kentrosaurus*. (b) *Stegosaurus*.

The distinctive armor of stegosaurs has long fueled speculation about its position, arrangement, and function. The spikes appear to have been primitively paired, though the 17 plates of *Stegosaurus* form a single alternating, nearly medial row. Sharpey's fibers, which reflect the direction of muscle attachment to bone, show that these plates stood upright, not flat, and could not have been moved up and down. Defense is a possible function, although it must be asked why stegosaurs lost all but their paravertebral row of armor. Stegosaurs were not fast runners: the lower limbs were long and the feet and upper limbs were short, so they would have had to stand and deliver. The spikes and plates, like the shoulder spines, could have had a passive function in defense, and the terminal tail spikes, which dragged along the ground, might have made a formidable weapon. However, the plates are relatively weak "sandwiches" of latticelike bone internally, not optimal for defense. In some large stegosaurs, the broad surface area of the plates may have contributed to thermoregulation, though this hypothesis has been challenged. However, it is most likely that they were used in species recognition: no two stegosaurs have identical armor.

Ankylosaurs

Ankylosaurs are distinctive dinosaurs (**Fig. 6**). Their skulls are short, low, and flat, always wider than high in rear view, and a complement of dermal ossicles covers the skull and closes the antorbital and mandibular fenestrae. There is an S-shaped row of small, uncomplicated teeth, which sometimes display wear. The body is broad and squat, and the limbs short. Armor covers much of the body, as in Thyreophora primitively, and appears in several shapes, including keeled or spined plates, polygonal pustules, spikes, and rows of symmetrical ovals or rounded rectangles.

Ankylosauria is divided into Nodosauridae and Ankylosauridae, based on a number of distinctive but subtle features of the skull and its plating, plus several postcranial characteristics. Several authors now recognize a third group, Polacanthidae. There are not many gross differences, though ankylosaurids have a tail club (four armor plates envelop the last series of tail vertebrae, which are deeply internested and partly fused). Both groups are primarily Cretaceous, though several

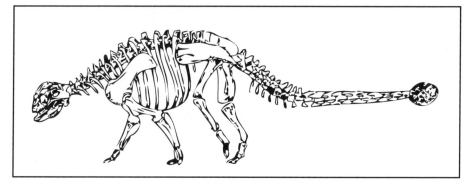

Fig. 6 *Euoplocephalus*, an ankylosaur.

genera are now known from the Jurassic. Ankylosaurs, like stegosaurs, seem never to have been very diverse, and they may have been solitary or have traveled in small family groups. They could not have been fast runners, and may have depended on camouflage, squatting, or slashing with lateral spines or tail club to repel predators.

CERAPODA: ORNITHOPODA

In many ways, Ornithopoda includes the most generalized ornithischians, which retain the bipedal posture of ancestral dinosaurs while lacking the horns, frills, spikes, and armor of other ornithischians. Nonetheless, they have their own specializations, and through the Mesozoic Era a general progression in size and adaptation can be seen from heterodontosaurs through hypsilophodonts to iguanodonts and hadrosaurs. In ornithopods, the upper front (premaxillary) teeth are set below the cheek (maxillary) teeth, and the jaw joint is set substantially below the occlusal plane of the teeth. *Heterodontosaurus* (**Fig. 4**), a fox-sized, lightly built form from the Early Jurassic of South Africa, shows these general characters and a few unusual ones, including protruding "canine" teeth presumably for defense or display.

The remaining three ornithopod groups are somewhat larger in size; they lose progressively all the front (premaxillary) teeth, reduce and/or lose the antorbital and maxillary fenestrae, and reduce the third finger (the fourth and fifth are already reduced in ornithischians). *Hypsilophodon* is a generalized member of this group. It still has five pairs of premaxillary teeth, and there is a gap between its front and cheek teeth, but the cheek teeth are now closely set and their crowns are typically worn flat. It was once thought that *Hypsilophodon* lived in trees, but this was based on the presumption that the first toe was reversed (as in birds) to serve as a perching adaptation. Actually, the anatomy of the foot is normal, and there are no additional reasons to put *Hypsilophodon* in trees.

More derived (advanced) ornithopods, such as *Dryosaurus* and *Tenontosaurus*, approach the condition of iguanodonts and hadrosaurs in many respects: the front teeth are completely lost; the front of the beak begins to flare; the cheek teeth become larger and more closely set, creating a single grinding battery of teeth; and the "bridge" of the nose (the area between the eyes and nostrils along the skull midline) becomes more highly arched. The forward prong of the pubis begins to grow outward until it reaches farther forward than the front of the ilium. As the cheek teeth begin to unify into opposing dental batteries, the internal kinesis (mobility) of the skull bones against each other also evolves to resist and redirect the forces of chewing.

The iguanodonts (**Fig. 7**) and hadrosaurs (**Fig. 8**) are the most familiar ornithopods, and were among the first dinosaurs discovered; they are still among the best represented in the fossil record. These two groups continue the trends already seen in the hypsilophodonts and dryosaurs, gaining larger size, losing the front teeth, flaring the snout further, and arranging the cheek teeth in a long, straight row for efficient crushing and slicing. In iguanodonts and hadrosaurs, the pubis changes somewhat: its original (backward) prong is reduced to a splint, and the forward prong enlarges and flares. This may reflect support for the intestinal tract, which is thought to have needed an extensive system of fermentation for the tough, low-nutrient plants that the animals ate. The claws on the fingers and toes are flattened and broadened into spoon-shaped hooves, and the middle three fingers, as with the middle three toes, may have functioned as a unit in walking on four legs. The fifth finger in these dinosaurs is quite distinctive: it points to the side, much like a thumb, and was similarly opposable to the other fingers, probably for manipulating food. The thumb is extraordinary in iguanodonts: the thumb phalanx, its terminal ungual, and its supporting metacarpal bone are fused together into a conelike spike, and this in turn is articulated tightly to the blocky, nearly immobile wrist bones. Whatever function this had in scratching, procuring food, or even defense, it had no function at all in hadrosaurs, which lost the thumb altogether.

Hadrosaurs (known as duckbills) dominated most of the Late Cretaceous of the Northern Hemisphere in numbers and diversity; their bones are found by the thousands in some deposits, suggesting mass deaths, perhaps during seasonal migrations or droughts. Their jaws were formidable food processors. Four or five rows of replacement teeth accompany each cheek tooth, and all of them are closely set, so as the teeth wore down they formed a single cutting and crushing surface of hundreds of teeth in each quadrant of the jaws. As in all cerapodans, the enamel was thicker on the oral side of the lower teeth and the cheek side of the upper teeth, so the teeth wore down unevenly: the cutting surface of the dental batteries was not horizontal, but beveled for better slicing as well as crushing.

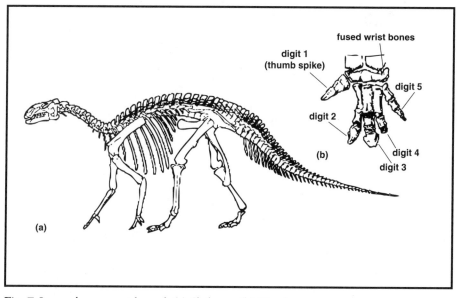

Fig. 7 *Iguanodon*, an ornithopod. (a) Skeleton. (b) Hand.

Hadrosaurs are generally divided into two groups (**Fig. 8**): the more conservative hadrosaurines and the more derived lambeosaurines, which are noted for their profusion of skull crests and ornaments. Some crests were solid and spikelike (*Saurolophus*), while others were partly hollow and resembled bizarre hats or helmets (*Lambeosaurus, Corythosaurus*) or even snorkels (*Parasaurolophus*). The partial hollowness in the crests, which were formed of the premaxillary and nasal bones, was to accommodate the nasal passages that normally run through these bones. Many explanations have been proposed for this great variety in skull structure, including functions in head butting, underwater feeding, or improvement in sense of smell, but most have little support. Because lambeosaurines vary little in other skull and postcranial features, a plausible explanation is that the crests were for display and species recognition, like the horns of present-day hoofed mammals. Experimental analysis and acoustical modeling of hollow-crested forms suggested that the chambers in the crests could have served as vocal resonators to create distinctive and far-reaching sounds for species recognition and communication. This idea has contributed to the picture of sophisticated social organization that is developing for many groups of dinosaurs.

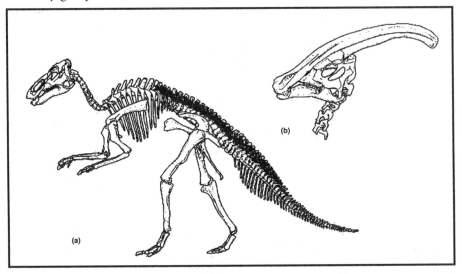

Fig. 8 Hadrosaurs. (a) Skeleton of the hadrosaurine *Edmontosaurus*. (b) Side view of skull of the lambeosaurine *Parasaurolophus*.

CERAPODA: MARGINOCEPHALIA

Marginocephalia includes the dome-headed Pachycephalosauria (**Fig. 9**) and the frilled, usually horned Ceratopsia (**Fig. 10**). Both groups are almost exclusively Cretaceous, though a few Late Jurassic forms are now known. Marginocephalia are recognized by a small shelf or incipient frill that overhangs the back of the head; also, their hips are wider than in most other forms, though ankylosaurs have extremely wide (but very different) hips. The small shelf on the skull was the basis for the elaboration of very distinct features and functions in both Pachycephalosauria and Ceratopsia. Known from a partial postcranial skeleton from the earliest Cretaceous of Germany, *Stenopelix* is possibly an early marginocephalian.

Pachycephalosauria

Pachycephalosaurs (**Fig. 9**) are distinguished by their thick skulls, which are often ornamented with knobs and spikes. Originally, the roof of the skull was rather flat but sloped upward posteriorly and was only slightly thickened (*Homalocephale*); the two upper temporal openings on top of the skull were still prominent. In more derived forms (*Prenocephale*, *Stegoceras*, *Pachycephalosaurus*), the skull roof became rounded into a dome about the shape of a human kneecap but much thicker, and the temporal openings were covered by bone. Rows and clusters of barnacle-shaped bony knobs adorn the nose, the temporal region, and the posterior shelf in most forms, and these can be developed into long spikes surrounding the dome (*Stygimoloch*). The thick, columnar bone that makes up the dome has often been interpreted as a shock absorber for head butting between competing males. However, more recent studies suggest that head-butting was quite unlikely in most forms, because the domes were so poorly designed for it. The rounded shape would cause strong torque on the neck; the dome may not have been mechanically strong enough for direct head impact; and early pachycephalosaurs had flat skull roofs that appear to have been more suited for pushing than ramming. The shape of the dome varied considerably in pachycephalosaurs, and so did its ornamentation; one pachycephalosaur had a dome with a sharp medial ridge and large spikes protruding from the sides of the head. Generally strong construction of the skull, the reinforced vertebral column (equipped with pencil-thick ossified tail tendons for support), and the strengthened pelvis suggest highly developed intraspecific (and possibly interspecific) use, perhaps mainly in merely displaying skulls to avoid direct combat.

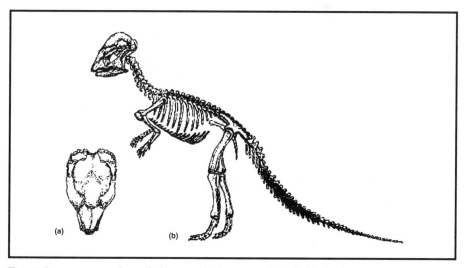

(a)

(b)

Fig. 9 *Stegoceras*, a pachycephalosaur. (a) Top view of skull. (b) Skeleton.

Ceratopsia

Ceratopsia (**Fig. 10**), along with the hadrosaurs, were the most successful Late Cretaceous ornithischians. They are distinguished by the presence of a rostral bone, a unique, toothless, neomorphic bone that caps the front of the upper beak much as the predentary caps the lower beak in ornithischians. Their earliest members, the basal (primitive) ceratopsians and psittacosaurs of the Late Jurassic and Early Cretaceous, were bipedal, slightly over 1 or 2 m in length (3–6 ft), and had a parrotlike beak; they were presumably browsers on low plants. The protoceratopsids initiated a trend to slightly larger size (2–3 m or 6–9 ft), quadrupedality, larger skulls, a more pronounced beak, a bony bump above the nose, and an enlarged frill at the back of the skull that usually appears as an upward-sloping arch formed by the squamosal and supporting bones.

Ceratopsidae, the great ceratopsians of the latest Cretaceous, include the centrosaurs and the chasmosaurs (**Fig. 11**). The centrosaurs generally had prominent nose horns but lacked horns above the eyes; the chasmosaurs emphasized the eye horns at the expense of the nose horns. Both groups evolved dental batteries of multiple rows of beveled teeth strikingly similar to the dentitions of hadrosaurs, and as in hadrosaurs the claws were flattened into spoon-shaped hooves. The frills were often ornamented with knobs and spikes and, as in hadrosaurs, are the most distinctive and differentiated parts of the body, the postcrania being rather conservative and not highly varied among ceratopsids. This also speaks to a function in display, rather than defense, for the great frills, which (with the notable exception of the short-frilled *Triceratops*, perhaps the best-known ceratopsian) were not solid

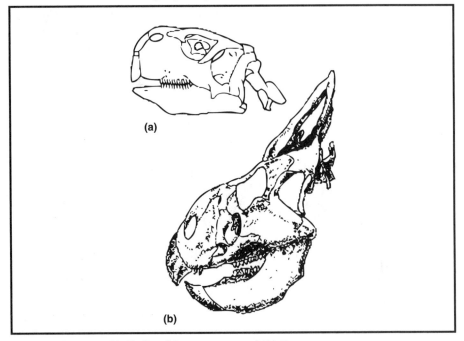

(a)

(b)

Fig. 10 Ceratopsia. (a) Skulls of *Psittacosaurus* and (b) *Protoceratops*.

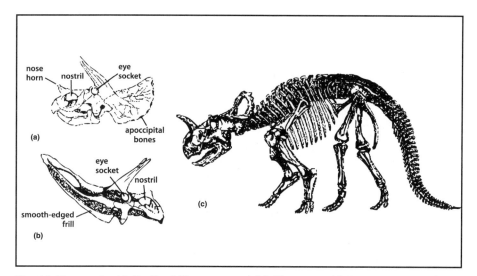

Fig. 11 Ceratopsia. (a) Skulls of *Triceratops* and (b) *Torosaurus*. (c) Reconstruction of *Centrosaurus*.

shields but merely bony arches—hardly a suitable defense against a marauding *Tyrannosaurus* or even another ceratopsid's horns. Species recognition is a more likely use.

SAURISCHIA

Saurischia (**Fig. 12**) includes Sauropodomorpha and Theropoda. For many years it was doubted that Saurischia formed a natural group because its two subgroups are so different, but in 1986 J. Gauthier demonstrated its validity cladistically. Uniquely derived features of Saurischia include the long neck vertebrae; the long asymmetrical hand in which the second digit is the longest; the slightly offset thumb with its short basal metacarpal, robust form, and large claw; and several other features of the skull and vertebrae. Unlike ornithischians, saurischians are fairly well represented in Late Triassic faunas as well as in the later Mesozoic.

Sauropodomorpha

This group includes the largest land animals of all time. The first sauropodomorphs (**Fig. 13**) elaborated several basic saurischian features: they had rather small skulls, elongated leaf-shaped, coarsely serrated teeth, long necks (with 10 vertebrae), robust limbs with a particularly robust thumb and claw, and a trunk that is longer than the hindlimb. *Thecodontosaurus* and *Massospondylus* typify this evolutionary stage and were probably mostly bipedal, with lengths of 2–6 m (6–18 ft). Soon, larger forms such as *Plateosaurus* appeared, and the trend toward larger size continued with the melanorosaurids of the Late Triassic and Early Jurassic. The necks became progressively longer, the bodies more massive (up to 10 m or 30 ft), and the limbs more robust, until with forms like *Vulcanodon* the condition seen in the great Sauropoda proper was attained. The relationships of these early forms,

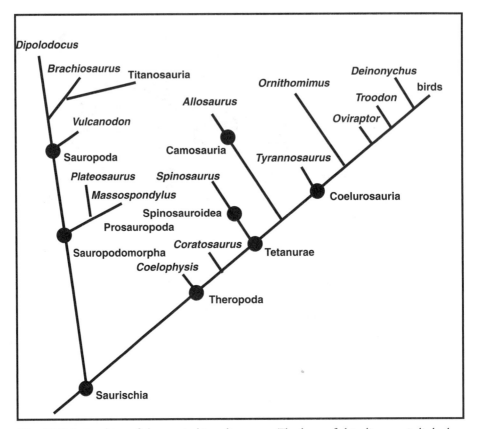

Fig. 12 Relationships of the saurischian dinosaurs. The base of this diagram is linked to Fig. 4. (*After theropod and sauropodomorph phylogenies, presented by various authors, in D. B. Weishampel et al., eds., The Dinosauria, 2d ed., University of California Press, 2004*)

collectively called Prosauropoda, are controversial—some authors regard them as a natural group, others suggest that they are ancestral to sauropods.

Sauropoda (**Fig. 14**) includes over 100 valid genera, which are split into a number of groups. A variety of primitive forms lived during the Late Triassic to Middle Jurassic; from the Middle Jurassic onward, the more advanced neosauropods appeared. Sauropods are, in general, much larger animals than prosauropods and have at least 12 neck vertebrae (with a complex series of bony laminae and struts) and stout, columnar limbs. Most of the familiar sauropod groups (for example, diplodocids, brachiosaurids, and titanosaurs) are neosauropods and can be distinguished from their more primitive relatives by a number of features, including an absence of tooth serrations and a reduction in the number of wrist bones. Each sauropod group has distinctive features. For example, the diplodocids had forward-sloping teeth similar to the prongs of an iron rake, very long necks, and long, whip-like tails. About halfway along the tail, the tail chevron bones beneath the vertebrae flatten into a canoelike shape that has been interpreted as a possible adaptation for rearing up on the back legs (tripodlike) to feed in higher branches. The brachiosaurs, by contrast, had tails that were shorter than their necks, and the tail lacked

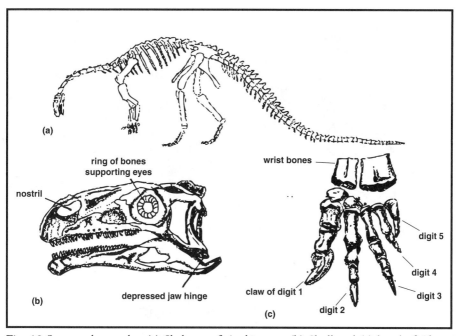

Fig. 13 Sauropodomorpha. (a) Skeleton of *Anchisaurus*. (b) Skull and (c) hand of *Plateosaurus*.

the canoe-shaped chevrons of diplodocids; but their forelimbs were longer than their hindlimbs, which perhaps compensated for this lack and elevated the chest region, allowing habitual high browsing. Some sauropods, notably the titanosaur *Argentinosaurus*, could reach an estimated 70 tonnes (or 77 tons) in weight; *Seismosaurus*, a diplodocid, may have been up to 30 m (90 ft) or more in length.

Theropoda

Although meat eating seems to have been the primitive habit of dinosaurs and their close relatives, the theropods are the only group of dinosaurs to remain carnivorous. (A very small number of theropods became herbivorous.) Historically, they were divided into large carnosaurs and small coelurosaurs, virtually for convenience; but in 1986, Gauthier showed that Theropoda is a natural group united by many features of the skull and vertebrae. In addition, the fourth and fifth fingers are lost or nearly so; the claws and teeth are sharp, and the teeth are recurved and serrated; the foot is reduced to the three middle toes; and the bones are lightly built and thin-walled.

Early in its history, Theropoda split into several lineages of primitive theropods (including coelophysoids and ceratosaurs) and a more advanced group, the Tetanurae. Primitive theropods, like the Late Triassic *Coelophysis*, were not large, but some, like the Early Jurassic *Dilophosaurus*, soon evolved to lengths of 5 m (15 ft) or more. Such ceratosaurs were the principal carnivores until the Late Jurassic. They were replaced by a great diversification of tetanuran theropods, which themselves split into three major groups: the Carnosauria (including *Allosaurus* and *Charcharodontosaurus*), the Spinosauroidea (such as *Spinosaurus* and *Baryonyx*),

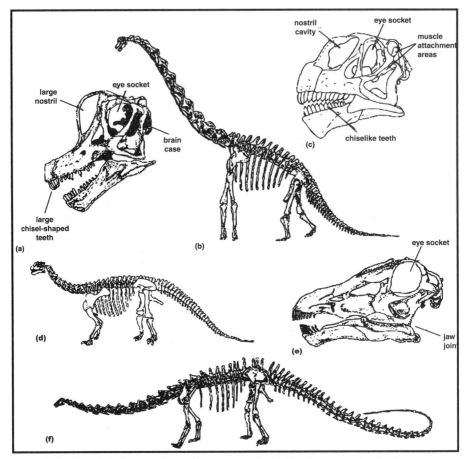

Fig. 14 Skulls and skeletal outlines of sauropods. (a, b) *Brachiosaurus*. (c, d) *Camarasaurus*. (e, f) *Diplodocus*.

and the Coelurosauria (including *Ornithomimus*, *Tyrannosaurus*, *Oviraptor*, *Deinonychus*, and the birds; **Figs. 15** and **16**).

Carnosaurs had large heads with enormous, daggerlike teeth. As in nearly all theropods, the teeth were laterally compressed, with keels fore and aft for slicing, and along each keel were fine serrations that improved tearing. There is some controversy about whether large theropods were predators or scavengers. Nearly any carnivore is opportunistic enough to scavenge, and no predator takes on the strongest members of a prey species: they prefer to prey on the old, the young, and the disabled. Hence it would seem very difficult to tell the difference between adaptations for predation and those for scavenging. At any rate, few animals would want to take their chances with *Allosaurus*. Spinosauroids were similar to carnosaurs in many respects, though they possessed a number of unique specializations. For example, many forms had elongate, crocodile-like snouts, and *Spinosaurus* had tall spines on its vertebrae that might have supported a large "sail" of skin or other tissue during life.

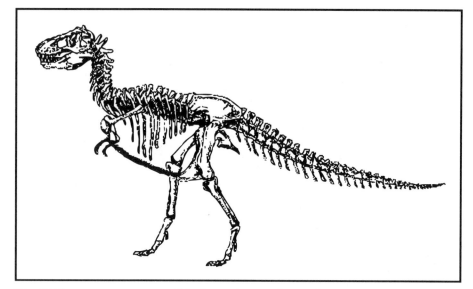

Fig. 15 *Tyrannosaurus rex*, one of the largest known theropods.

The coelurosaurs were highly varied dinosaurs. The ornithomimids, or "ostrich dinosaurs," had reduced (or no) teeth, very long legs, and a general build reminiscent of the largest birds of today. Oviraptorosaurs had bizarrely fenestrated skulls and toothless beaks, yet the hands had sharp claws. Troodontids had long snouts with small, sharp teeth, their brains were large, and their legs and feet long. Tyrannosaurids were gigantic superpredators with massive skulls and strongly reduced forelimbs (with only two fingers per hand). The dromaeosaurs, which included the well-known *Deinonychus* and the "raptors" of the film *Jurassic Park*, had relatively large skulls and teeth, long prehensile hands and clawed fingers, a stiffened tail, and one other feature also found in some troodontids: the second (inside) toe on the foot was held off the ground and sported a trenchant, enlarged, recurved claw. In 1969 it was hypothesized that these animals would have used their long arms to grab prey, their tails to help them balance, and their large claws to rip open the bodies of their quarry. Only a few years later the Polish-Mongolian expeditions returned from the field with a pair of specimens that supported the 1969 prediction: a *Velociraptor* with its hands on the frill of a *Protoceratops*, and its feet at the prey's belly.

Dromaeosaurs were also the closest known forms to birds. The origin of birds from within dinosaurs is old: T. H. Huxley noted the strong resemblances in the 1860s. A century later J. Ostrom recognized the many unique similarities shared by *Archaeopteryx*, the earliest known bird (**Fig. 17**), and coelurosaurs, particularly forms such as *Compsognathus* and *Deinonychus*. The pubis had begun to project backward; the arms were long; the hands and fingers very long; and their digits correspond to those in the bird hand. A bony sternum had been found in some forms, and also the fused clavicles that in birds form the wishbone. The foot was three-toed, and the first toe rotated backward as in birds. The problem was ana-

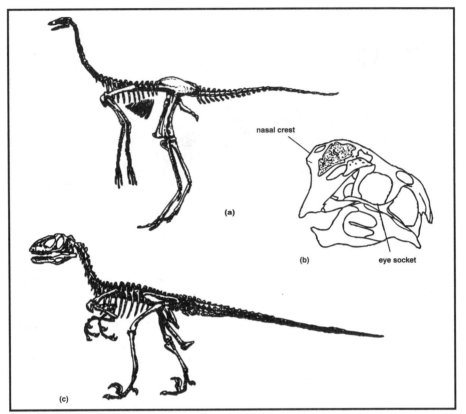

Fig. 16 Coelurosauria. (a) Ostrich dinosaur, *Ornithomimus*. (b) Skull of *Oviraptor*. (c) Skeleton of the dromaeosaur *Deinonychus*.

lyzed cladistically, and it was concluded that nearly 200 shared derived characters placed birds squarely within the coelurosaurian dinosaurs. In evolutionary terms, then, birds are dinosaurs, so at least one lineage of dinosaurs is not extinct.

PALEOBIOLOGY

Dinosaurs laid eggs, and many nests have been found (**Fig. 18**), but matching nests and eggs to their makers is difficult unless embryos are preserved. Even then, juveniles usually lack many diagnostic characteristics of adults, so precise identification is difficult. Embryos of the hadrosaur *Maiasaura*, found in Late Cretaceous deposits from Montana, had uncalcified bones with cartilaginous joint ends, and the hatchlings were clearly helpless. Their eggshells were crushed into fragments, and the skeletons of the hatchlings found in the nests were clearly too large to have recently emerged. It appears that the young stayed around the nest and were fed by the parents until their bones fully calcified and they could fend for themselves. It is not clear how widespread either of these behaviors (or others) was among dinosaurs, but birds (living dinosaurs) are known for their extended parental care,

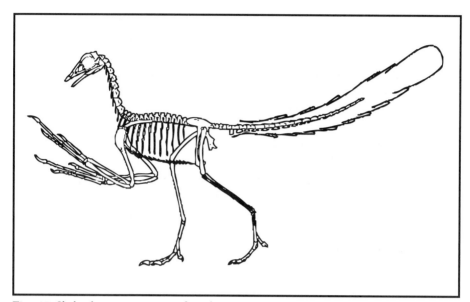

Fig. 17 Skeletal reconstruction of *Archaeopteryx*, the earliest known bird, a saurischian theropod.

and crocodiles, their closest living relatives, also care for their young, so it may be a general behavior among archosaurs.

GROWTH

If their bones are any indication, dinosaurs grew rapidly. Microscopic studies of their long bones show that the tissue was well vascularized as in birds and mammals, and that it was rapidly replaced. The texture of the bone is predominantly fibrolamellar, as in birds and mammals, and juvenile bone also shows a highly woven pattern reflecting rapid growth. Rest lines of regular occurrence have been reported in most dinosaur bones, as in those of many birds and mammals, and like these they appear to be annual. The amount of bone tissue and its many large blood vessels indicate that a lot of bone could be deposited annually, especially in juveniles and in large species. Tyrannosaurs, for example, grew twice as fast as African elephants to the same size. When dinosaurs reached full size and sexual maturity (perhaps not simultaneously), several bones co-ossified (for example, the bones that make up the braincase), as in birds and mammals. However, most dinosaur skeletons do not show these features, so they may not be fully adult.

SOCIAL ORGANIZATION

Footprints provide strong evidence that many kinds of dinosaurs traveled in groups, at least occasionally, and this is supported by records of mass burials in

Coelophysis, *Plateosaurus*, *Allosaurus*, *Centrosaurus*, and various hadrosaurs. Social organization has the benefits of protection for the young and weak and the communication of information about food and other resources. Most major groups of vertebrates and even some invertebrates have some complex forms of social organization. Whether traveling in family groups or (at least occasionally) in large herds, all dinosaurs of any appreciable size would have had to migrate to exploit food sources successfully, if only in a local area. Annual migrations may have followed seasonal patterns of weather and vegetation.

A general picture of dinosaurian social behavior—no doubt highly variable among groups—is drawn from inferences about mass remains of trackways and bones, parental care, and the great variety of skeletal features conspicuously related to intraspecific interactions. The diversity of horns, crests, domes, knobs, and frills in many dinosaur groups contrasts starkly with the relative uniformity of their postcranial skeletons. Selection for these ornaments cannot be explained in purely functional or random terms, but strong analogies can be found in the rather similar features of living birds and mammals. Communicating, recognizing members of the same species, attracting mates, and repelling rivals, as well as delivering similar signals to members of different species, are all behavioral functions of such structures in living animals.

METABOLISM

Dinosaurs evolved from within Reptilia (and so, therefore, did birds), but they are as unlike living reptiles as bats and whales are unlike horses among living mammals (and they may have been just as diverse metabolically). The physiology of extinct animals can be assessed only indirectly. No one doubts that a fossil horse was a warm-blooded, endothermic homeotherm, but such statements cannot be made about the earliest mammals and their ancestors. In the same way, the evolution of thermal strategies in dinosaurs was probably mosaic, depending on the adaptations of individual groups, and should not be considered an all-or-nothing proposition of hot-bloodedness. Many of the lines of evidence discussed above suggest that

Fig. 18 Nest of ornithischian dinosaur eggs from Montana.

Mesozoic dinosaurs were similar in behavior and activity to mammals and birds; no evidence seems to ally them physiologically to crocodiles and lizards. But dinosaurs should be taken on their own terms, not shoehorned into models of reptile or mammal physiology based on available living analogs. The recent discoveries of nonavian dinosaurs with feathers and featherlike coverings from Early Cretaceous deposits in China indicate that the thermoregulation of at least some theropods may have been birdlike, as bone growth also shows.

DIVERSITY

For 160 million years during the Mesozoic Era, dinosaurs diversified into between 500 and 1000 known genera, though many of these are based on single (often partial) specimens. Through time, dinosaurian diversity increased, but the difference between preserved diversity and some estimates of projected diversity is traced in part to varying availability of the rock record plus differential exploration. Evolution in dinosaurs was rapid: few dinosaurian genera survive more than the temporal span of a typical geologic formation (a few million years), but close relatives are often observed in succeeding formations. At the beginning of the Age of Dinosaurs, the continents were just beginning to drift apart. Dinosaurs are known from every continent and are often used to suggest land connections, such as between North America and Asia in the Late Cretaceous, or the isolation of South America during much of the Cretaceous. Some groups, such as the Ceratopsidae, apparently radiated in isolation on one or another continent.

EXTINCTION

Many causes have been proposed for dinosaur extinction at the end of the Mesozoic Era, but most (for example, reproductive sterility, plant toxin poisoning, cataracts, supernova explosion, glaciation) have no supporting evidence. The apparent sudden decline of dinosaurs is surprising in view of their long dominance. However, the Late Cretaceous was globally unsettled in many ways. Great volcanoes were exploding in India; the Rocky Mountains were beginning to form; successive waves of regression and extinction had rocked the world's oceans; and the shallow seas were draining from many continents, including North America and Europe. These changes made the climate on land less equable and more seasonal, paving the way for destabilization of ecosystems, and perhaps creating patchy environments in which new plants and animal associations flourished. In the last few million years of the Cretaceous, known dinosaur diversity, based mainly on excellent exposures from the United States western interior, declined precipitously until, in the meters of sediment just below the Cretaceous-Tertiary boundary, only *Triceratops* and *Tyrannosaurus* survive.

An explanation is needed for this drop in diversity. Macroevolutionary patterns are based on the origination of new species and the extinction of existing ones. Dinosaur taxa were of short duration and rapid turnover; thus extinction was ubiquitous and quick throughout dinosaurian history. Toward the end of the Cretaceous, dinosaurian originations were at first burgeoning, but then they dropped; extinctions were proceeding at a near-normal pace and eventually overtook the origination rate until diversity fell to only a few taxa. In 1980, it was proposed that a giant asteroid had struck the Earth at the end of the Cretaceous, sending up a cloud of dust and vapor that blocked the sunlight for a few months or years, first lowering and then raising temperatures and causing widespread marine and terrestrial extinctions. These effects were later augmented by scenarios of global wildfires, intensely acidic rain, and other proposed environmental poisonings. As valid as the proposal of a giant asteroid impact now appears, some of the proposed biotic effects are clearly overdrawn. Dinosaurs were already declining in diversity by that time. The birds survived the impact's effects, and so did many groups of fishes, sharks, amphibians, lizards, snakes, crocodiles, and mammals. Many of these groups are notoriously sensitive to environmental disturbances. Hence any catastrophic scenarios for the terrestrial biota at the end of the Mesozoic must account both for the latest Cretaceous decline in dinosaurian diversity and the survival of any proposed environmental catastrophes by other terrestrial animals and plants.

SOURCES

R. T. Bakker and P. M. Galton, Dinosaur monophyly and a new class of vertebrates, *Nature*, 248:168–172, 1974

P. J. Currie and K. Padian (eds.), *The Encyclopedia of Dinosaurs*, Academic Press, 1997

J. O. Farlow and M. K. Brett-Surman (eds.), *The Complete Dinosaur*, Indiana University Press, 1997

J. A. Gauthier, Saurischian monophyly and the origin of birds, *Mem. Calif. Acad. Sci.*, 8:1–55, 1986

K. F. Hirsch, K. Carpenter, and J. R. Horner (eds.), *Dinosaur Eggs and Babies*, Cambridge University Press, 1994

P. C. Sereno, The evolution of dinosaurs, *Science*, 284:2137–2147, 1999

D. B. Weishampel, P. Dodson, and H. Osmolska (eds.), *The Dinosauria*, 2d ed., University of California Press, 2004

ADDITIONAL READINGS

J. R. Horner, Dinosaur reproduction and parenting, Ann. Rev. *Earth Planetary Sci.*, 28:19–45, 2000

M. G. Lockley, *Tracking Dinosaurs: A New Look at An Ancient World*, Cambridge University Press, 1991

Dinosaur Dreams[*]

Reading the Bones of America's Psychic Mascot

By Jack Hitt
Harper's Magazine, October 2001

Sixty-five million years ago, conservatively speaking, the last dinosaurs lay down and died—on the ground, beside rivers, in tar pits. Then, about a hundred years ago, they got back up and have been pretty busy ever since. Hardly a week goes by that they don't make the news, because of a new theory about either how they lived or how they died. There might be word of a new prime-time TV deal, another revelry (Dinofest V is planned for next year), a new exhibit, the goings-on of paleontological hunk Paul Sereno, a Spielberg script, a hot toy, a legal dispute about some bones, an egg.

In America, where dinosaurs do most of their work (and always have), they periodically disappear from view and then resurface, like John Travolta or Democrats, capturing and losing our cycling interest. Dinosaurs are distinctly American, not only because our scholars have so often been at the forefront of fossil discoveries and paleontological theory but because the popular dinosaur is a wholly owned projection of the nationalist psyche of the United States. Their periodic rebirth in pop culture neatly signals deep tectonic shifts in our sense of ourself as a country. Even glancing appearances can be telling. After Newt Gingrich rampaged through the House of Representatives and seized power in 1994, he placed a skull of T. rex in his office. When the current resident of the White House got a new dog, he called him, strangely, "Barney"—a name most closely associated with a phony dinosaur who masks his cheerful dimness with sticky compassion.

It's been almost exactly a century since the first one stood up in New York City amid a Miramaxian media circus, and, bookending nicely, we seem to be in the clutches of another rage. One recent estimate asserts that there are probably more dinosaurs on the earth now than there were in the Mesozoic Era. The current rate of finding new species is one every seven weeks. Half of all the known dinosaurs have been discovered in the last few decades. What really marks our era are the new tools—amino-acid mapping, the CAT scan, treadmill-energetics studies—that permit us once again to read meaning in the bones and to imagine the world that was when they held up flesh.

Determining the causes of America's periodic obsession with dinosaurs is tricky. They have done a lot of heavy lifting, culturewise. Besides in children's narratives (where dinosaurs still rule the world), they have served as political totems, deranged kitsch, icons of domesticated terror, cultural mules for Darwin's (still) troubling theory, and environmental Cassandras resurrected to act out her famous final words, "I will endure to die."

In the pop-culture chaos that is uniquely our own and that swirls around paleontology's quiet drudgery, certain dinosaurs burble up to the top of our celebrity-oriented mass-media-marketplace mosh pit to float about in plain view, gazed upon by everyone. And we the people elevate them out of that frenzy as surely and by the same circuitously collective route as we do Ricky Martin, Stone Phillips, Cameron Diaz, or Tom Wolfe. Going in and out of fashion, these giants are creations of our own making as much as the paleontologists', perhaps more so. In demanding to see them, we sculpt their meaning; like outsized Schrodinger's cats, their existence depends on whether or not we have decided to look at them.

None of these gargantuans has been more gawked at than *Tyrannosaurus rex*, the Mick Jagger of dinosaurs. And yet he's been absent on the national stage lately (as if he were off in rehab plotting his comeback tour), and there's a reason. Controversial scholarship has turned up a new interpretation of how the great meat-eater lived, and it is so at odds with T. rex's public persona that even scholars hate to talk about him anymore. In a sense, the scientific reality of the King of the Tyrant Lizards has laid bare our symbolic uses of him. So T. is in hiding.

Developed by the self-credentialed iconoclast Jack Horner, the theory holds that T. was not the great predator who marauded through primordial landscapes but rather a slow, putzy scavenger that poked around the Cretaceous countryside in search of maggoty carrion.

Horner can mount a great deal of technical evidence to prove that T. was no hunter, and I have seen him turn an audience completely around in an hour. T. had densely muscled legs with calves the length of his thighs, which are good for walking but lousy for pouncing. T. had poor eyesight, not a great trait, since hunters typically track prey at twilight. CAT scans of T.'s skulls have revealed massively dominant olfactories; he could smell from as far away as fifty to seventy miles, a handy adaptation if the meat you're looking for is malodorously ripe.

Famous predators, like lions or velociraptors, have powerful forearms to catch their lunch or to hold their prey while they rip fresh muscle from the bone with their backlegs. T.'s arms were thalidomidal, about as helpful as having two fussy little hands growing out of your nipples. "T. rex couldn't even clap," Horner points out.

T. didn't walk the way every comic book and *National Geographic* magazine used to show—upright, head raised, front claws poised—rampant, as heraldic buffs say.

None of the preserved trackways showing the footprints of theropods (T.'s family) have an impression of a dragging tail between them. Rather T. probably waddled, like the ten-ton vulture he was, tail straight out, body parallel to the ground like a . . . I believe the heraldic term is wuss.

Horner has nothing but critics, and they resist his logic with the willful stubbornness of biblical creationists weighing the merits of evolutionary theory. They will admit that T. ambulated like a monstrous sandpiper, but they insist he was still a predator, dammit. Even though many museums have reset T. in his new posture, they typically turn the head and open the jaws, as if he were just glancing at you while racing by en route to some old-fashioned predatory mayhem.[1] Unlike Cretaceous Era dinosaurs, twentieth-century ones do not die easily.

If you ask Horner why early curators first set T. in a predator's stance, he'll tell you it was because the "ceilings in museums were too high and they had to fill the space." To prove his point, he'll betray an embarrassing secret: the early curators had to smash and whittle T. rex's bones and then remove vertebrae to assemble him into that fighting posture. "There was more Barnum than science in those earliest displays," Horner says. "The curators realized it was a spectacle for the nation, and that's why T. rex looked the way he did. It was what the country wanted to see."

Dinosaurs had been discovered long before modern America took up the cause, but they were easily incorporated into nearby myth. Native Americans assumed the large bones belonged to the "father of buffaloes." In England a 1677 discovery was identified as a "human thigh bone of one of the giants mentioned in the Bible." A fossilized trackway in Connecticut of a giant three-toed creature had long been thought by locals to be that of the avian god Thunderbird until more advanced colonials corrected them to understand that it was the footprint of the "raven of Noah."

It was an English scientist who first announced in 1841 that these giant bone discoveries belonged to some new order of creature he decided to call "dinosauria." And the first full dinosaur, the iguanodon, was an entirely British discovery. But those early dinosaurs say more about the Old World's pinched imagination than about its paleontology. Iguanodon refers to a reptile's tooth, and the image the English conjured from those bones was nothing more than an obese crocodile on four piano legs. Such are the rhinocerine quadrupeds that populate the first dino narrative, Sir Arthur Conan Doyle's *The Lost World*. Until Darwin, one wasn't able, intellectually, to describe a world that didn't already exist. The geological *Weltanschauung* was biblical and stagnant, yet almost immediately it was clear that we had discovered a potent new metaphor.

The first big show of dinosaurs opened in England in 1854. To inaugurate the spectacle, the officials risibly set the dinner table inside the reconstructed body of an iguanodon. Later a famous French feast to celebrate the arrival of an American dinosaur included "*hors d'oeuvres paleontologiques*" and "*potage bisque aux Eryon jurassiclues*." We eat dinosaurs and they eat us. We partake of their dinosaurness, they partake of our humanness. From the beginning there was a commingling, something vaguely divine. As with Christian communion, we acknowledged our desire to become them by dining on them while being consumed by them. Theophagy is not a notion that casually erupts in a culture, even kitschily.

Yet the full promise of those early dinosaurs would require New World vision. In 1858 a professor of anatomy at the University of Pennsylvania named Joseph Leidy examined an early hadrosaurus discovered in New Jersey and came to the revolutionary insight that this monster didn't tread the ground like some brute from the bestiary but stood up. Leidy was an early pioneer of the new American field "natural history"—the very phrase overturning the biblical view of the earth as static with a rather New World ideal of looking at nature as an unfolding story, a "history."

Dinomania's great awakening came at the end of the nineteenth century, and it ultimately took the form of a competition—the Bone Wars—between Professor O.C. Marsh of Yale and Edward Cope of the Philadelphia Academy of Natural Sciences. They competed in getting bigger grants to mount large-scale expeditions to increasingly more remote outbacks—in order to bribe locals and hire minions (and eventually spies) to find ever bigger bones and inscribe more and more names in the permanent book of Latinate taxonyms.

These two created an inner tension in paleontology that continues today. Marsh was an establishment figure who easily ascended into the aerie of academe by becoming a professor at Yale. (His rich uncle George Peabody donated the nest egg for the New Haven museum that still carries his name and houses many of the most famous dinosaur skeletons.) Cope, on the other hand, was a self-educated man who bankrupted himself to finance his excursions. A volatile, excitable, roaming character, Cope was never comfortable with the institutional positions offered him. In the end, his suspicions of the highly connected world in which Marsh thrived drove him to James Gordon Bennett's *New York Herald*, where he initiated one of the most vicious personal attacks in history. He accused Marsh of destroying bones in order to shore up his reputation, of plagiarizing Cope's intellectual work, and of stealing and lying.

It may almost be slapstick legend at this point, but the feud was said to have begun early on in their careers when Marsh publicly noted that Cope had placed the head of a plesiosaur on—is there any nice way to tell somebody this?—the wrong end. (Given the contingent and fragmentary nature of the field, getting the bones wrong is naturally a frequent charge and almost always correct.) From that time on, dinosaurology has been populated by both Marshite establishmentarians, who strain to lend the discipline the donnish solemnity of the Old World trivium, and Copean renegades, who invigorate it with the improvisational air of the American hobbyist.

Today's Marshites are institutionalists like John Ostrom of Yale or even general paleontologists such as Stephen Jay Gould of Harvard. Copeans (often feeling aggrieved and carrying their flamboyant reputations to faraway institutes) range from autodidact Jack Horner (with the Museum of the Rockies in Montana) to the now deceased ex-maintenance-man-cum-world-renowned eggshell-expert Karl Hirsch to the well-publicized crank Robert Bakker, a ponytailed paleontologist with a penchant for gaudy Stetsons who was once "unpaid adjunct curator" at the University of Colorado Museum.

The Marshites and the Copeans need each other, though—the former for legitimizing a discipline populated by weekend enthusiasts and the latter for bringing adventure and creativity to a field that could easily settle into its own self-contented dust. These two characters are themselves deeply American types—the fugitive genius at the frontier goddamning all tradition and the starchy conformist trying to reconcile a voluble present with a soothing knowable past. It's what gives the stately field its inescapably puerile nature. Although "paleontology" is a word of scientific distinction, the taxon "dinosauria" means "terrible lizard," a phrase straight from the imagination of a five-year-old boy.

The first dinosaur bone ever identified was a huge double ball joint discovered in England in 1677. When it was illustrated in 1763, the artist Richard Brookes noted its resemblance to a monstrous pair of testicles. Brookes may have thought it belonged to a well-endowed biblical giant, but he may just have been goofing around when he named it. Marshites don't like to tell this story. But, as W.J.T. Mitchell argues in *The Last Dinosaur Book*, since it is the first bone ever identified, "by the strictest rules of biological nomenclature, *Scrotum humanum* is the true name of the dinosaur."

Scrotum humanum. Dinosaurology has never been able to quite shake its Jim Carreyness, which may explain why the field attracts so many kids and amateurs. Dinosaur theorists don't need advanced degrees in organic chemistry or technical prowess with superaccelerators—just a willingness to master the quite knowable bank of dinosaur findings, a chore that often begins just before first grade. The professional work can begin shortly thereafter. Three years ago, a dinosaur expert in New Mexico discovered a new creature in the Moreno hills and wrote an award-winning book about it—*Zuniceratops christopheri*, named for Christopher Wolfe, who was eight years old at the time.

Because just what is the intellectual task of paleontologists young and old? To take a few toylike objects—really big, really cool bones—and to imagine an entire world. It's a kind of intellectual play that is never troubled by comparison with any rigorous empirical reality. The only competition is some other imagined world that seems that much more neato.

In the first decade of the twentieth century, people everywhere were hungry to see the first inhabitants of this world as they emerged from the mythic frontier of the American West. Unlike England's, America's dinosaurs were easy to find. The nonacidic prairie soils of the West coupled with constant wind erosions and mild rains meant that at those early dinosaur digs, like Como Bluff and Bone Cabin Quarry, skeletons were often just lying exposed on the ground. The first dinosaur to go on tour was a plaster cast of a diplodocus, one of the long-necked sauropods that would eventually become world famous by the name of brontosaurus. For those willing to look closely, the connection between emerging American power and the vigor extant in the skeleton was apparent. That traveling dinosaur and a subsequent discovery were known, scientifically, as *Diplodocus carnegii* and *Apatosaurus louisae*, after their patrons, Andrew and Louise Carnegie.

New York's American Museum of Natural History unveiled the first permanent display: a brontosaurus in 1905, paired in 1910 with a T. rex. Figuring out how to hold up, say, T. rex's 2,000-pound pelvis was a chore. But as luck would have it, the industrial revolution was outfitting every American home with the new-fangled marvel of indoor plumbing. And that's what the curators chose—the same L joints, sink traps, U brackets, and threaded pipes that forged the infrastructure of America's emerging empire also cantilevered the spines, jaws, breastplates, and hipbones of those two great beasts.[2] It was certainly as much a celebration of our new power as it was of the dinosaurs, and right away they assumed oddly familiar personalities. The brontosaurus was a long-necked galoot—a cud-chewing, vegetarian, gentle giant. Then just across the aisle, the psychic opposite: T. rex, frenzied carnivorous killer. An interesting pair those two, and it is no coincidence that their erection occurred just before we entered World War I, revealing to the world the character of a new global species—the American: A big, dumb rube, until provoked—then berserker rage.

It is difficult to imagine the effect those first displays must have had on the minds of our great-grandparents. But consider: You had to make a big trip to New York to see them. Newspaper descriptions and the occasional picture only stoked the desire to go. Meanwhile, all around you, a greatness was coming together—electrical wiring, indoor plumbing, the plane, the car, the movie—and it was being assembled around, over, and through you into a colossus larger than anything since Rome. The emotion that surged when you tilted back your head to look at those early dinosaurs was awe, for sure, but it was also a suffused patriotism. The skeletons gave substance and turgor to a novel feeling of giantness that citizens must have felt as they sensed their own inexorable participation in a new American project of pure immensity, an awareness that something dinosaurian in scope was rising up in the world: The first modern superpower.

When the Roaring Twenties were in full swing, adventurer/dino hunter Roy Chapman Andrews of the American Museum of Natural History (said to be the model for Indiana Jones) set off for the Gobi Desert in search of bones. His expedition deployed America's newest projection of twentieth-century power, a fleet of cars. It's not altogether clear that Chapman was being ironic when he announced that the purpose of his trip (paid for by the new Dodge Motor Company) was "the new conquest of Central Asia."

With the rise of Nazism, our biggest dinosaur skeletons were considered matters of—and never has this phrase sounded so vulnerable—national security. As a precaution, New York's T. rex was clandestinely removed to a safe location: Pittsburgh, a place where even victorious Nazis might never go.

No sooner had the boys come home in 1945 than our Mesozoic doppelgänger began shouldering a new burden, the A-bomb. American power suddenly seemed "terrible" in the "dino" sense—indiscriminate, Oppenheimerish, annihilating. In Japan the pure destructive power of Hiroshima arose chthonically as the incarnation of American evil. Godzilla: T. rex with a foreign policy (and big forearms). In Hollywood, though, the postwar dinosaur was a means to explore science's limits. It all began with *The Beast from 20,000 Fathoms*, which begat *King Dinosaur*, *The Beast of Hollow Mountain*, *The Land Unknown*, *Gorgo*, and *The Lost World*, among others.

In *20,000 Fathoms* an atomic test at the North Pole revives a hybridized T. rex (again, big forearms) hell-bent on returning to the place of his birth, the Hudson River valley—a.k.a. New York City (a story line more deeply true than Hollywood would ever know). At last the authorities trap the monster at Coney Island, foreshadowing by half a century *Jurassic Park*'s melding of dinosaur and entertainment. As the monster ravages his way through the Tilt-a-Whirl ride, a sharpshooter hits him in the heart with a bullet packed with "radioactive isotope"—the very substance that revived him and now destroys him. The message of all these movies was clear: We can re-animate the monsters or restore them to the fossil bed. Hiroshima was terrible but containable. Behold, all ye who tremble, the majesty of American might and the righteousness of her restraint.

In the 1950s, as GIs took brides and settled into Levittowns everywhere, the dinosaur, too, was dramatically domesticated. How? One word, Benjamin: plastics. With the invention of this new pliable material, anyone could mold a dinosaur into any imaginable position. Plastic re-animated the dinosaur by putting a kind of flesh back onto those big immobile bones and giving each of them a smooth synthetic flexibility. This dinosaur stood up nearly lifesize at every Sinclair gas station, and miniatures populated the toy sections of the department stores. Severing its umbilical connection to the sober conservatism of paleontology, the dinosaur now entered pop culture as a free agent. This was one for the people, the masses— the democrasaurus. The late fifties saw the earliest dinosaur trading cards, the first dinosaur stamp, an explosion of comic books, and welcomed the sixties with *The Flintstones*.

Plastic was also the unique innovation of postindustrial America, ultimately replacing Mr. Carnegie's steel in our infrastructure and ushering in our technological revolution. Today, scarcely an item on the planet doesn't contain some. If dinosaurs ruled the earth once, plastic comes close to making that claim today. And what is plastic made from? The resurrection couldn't get more literal.

Like everybody else, dinosaurs turned off and dropped out in the sixties and seventies. There were some minor shifts. It was in this era that Horner developed his other controversial theory: that dinosaurs parented. Horner is almost singlehandedly responsible for getting dinosaurs in touch with their feminine side. He had discovered some dinosaur eggs and hatchlings surrounded by fossilized "regurgitated food" and adult dung—the detritus of a mother animal caring for its helpless nest-bound offspring. Dinosaurs had come a long way from the macho "terrible lizards" that erupted ab ovo, ready to begin their rampages. One of Horner's discoveries was *Maiasaura* (the first use of the feminine "a" ending versus the masculine "us"), meaning "good mother lizard." Around this time, in case you don't recall, the hottest talk-show host was Phil Donahue and the president was Jimmy Carter.

Dinosaurs' next epic pop-cultural leap in the national consciousness was the movie *Jurassic Park*. Like every curator before him, Spielberg sensed it was once again time to rearrange the Rorschachian bones. The star of that movie was the velociraptor. The brontosaurus and the T. rex make almost cameo appearances, like Robert Mitchum in the *Cape Fear* remake—an insider's nod to the grizzled original. The brontosaurus has only one significant scene, sneezing a few gallons of dino snot all over the kids. Still the goofball, after all these aeons. T. rex assumes his familiar role as enraged killer, but he's still a patriot, arriving at the end to save the day, like the cavalry.

And just who was the velociraptor in this 1993 movie? For a dinosaur, he was small, human-sized, and warm-blooded. The scientist in the film noted its jugular instincts ("lethal at eight months"), cunning ("problem-solving intelligence"), and strategic adaptability ("they remember"). At a time when the Japanese seemed to be taking over the world, we gazed upon a new beast—the global-business warrior, physically downsized, entrepreneurially fleet, rapaciously alert, ready for the dissolution of the nation-state. If the Pacific Rim was poised to take over the world, then this dinosaur was our response; an image that reflected how we conceived of our enemy as much as how we conceived of our new selves. The old-economy capitalist (T. rex) is there but sidelined, yielding to the distinctive features of the new-economy capitalist—lean, mean, smart, fast, and fatal. Many of the incidental descriptions of *Jurassic Park*'s velociraptor could easily be dropped into a *Fortune* magazine profile of Henry Kravis, "Chainsaw" Al Dunlop, or Larry Ellison, without any editing. Even its Latinate name eerily foreshadowed its future metaphorical role. "Velociraptor" means fast-footed thief.

After *Jurassic Park*'s success, the American Museum of Natural History shook up the bones once again and then, in 1995, tossed them out into a new display that received widespread praise. What the visitor sees is very much a new world, fully reimagined as a time of environmental balance. These dinos are shown in familial clusters, in mini-dioramas under glass. Many are small and seem as approachable in their outsized terrariums as gerbils in a suburban den. They are poised not to kill but to mate and to remind us, as the Disney movies do, of the Great Circle of Life. One display comes across like an intact family brimming with centrist heterosexual values—a mom and pop psittacosauri, plus three little hatchling psitts, gathered together, possibly on their way to church. A rare dinosaur fetus is also on display. The sum total points toward a very Al Gore-like dinosaur—the ecosaur.

At least, I think that's what it means. In these tenuous epistemological days, the museum adopts an unnerving tone. Each display states a bit of current dino theory and then mercilessly undercuts itself. One glass case explains the latest nasal theory regarding duckbills, but then beside the text a yellow warning label reads: "These are all intriguing hypotheses, but the fossils do not give us enough evidence to test whether any of them are correct. The mystery remains unresolved." That's a bit too much postmodern uncertainty to hang at the eye level of a ten-year-old, don't you think?

When I asked a museum official what he thought the entirety of such an exhibit added up to, he said, "Dinosaurs were the most successful life form that ever lived on this planet, and they became extinct. Extinction is a real part of life, and it's not so bad. When the dinosaur died out, the world went on and other species were created. One of those species was the human form. I think that, in all likelihood, our species will become extinct, and when that happens, that's probably not a bad thing."

So let me get this straight. We don't know anything and we're doomed. What a distance we've traveled since we looked at that first *Scrotum humanum* and saw our own lusty selves. Ecosaur doesn't begin to capture the sagging confidence and fear of empire-wide failure embodied here. Let's upgrade. Apocalyptosaurus.

Fortunately, it wasn't long before the osteo-reply to apocalyptosaurus arrived. In the winter of 1999, the National Geographic Society announced a "feathered dinosaur" exhibit with fresh specimens from China. The entire display represented a Tony Bennett-like revival for Yale paleontologist John Ostrom, whose brilliant hypothesis about the fate of dinosaurs had gotten obscured in the last two decades of fervent debate about extinction.

Among dozens of theories about the dinosaurs' fate—including global warming, subterranean-gas leaks, magnetic fields, trans-species miscegenation, and, of course, the meteor from outer space—Ostrom's idea suggests that some dinosaurs may have just evolved. Their streamlined progeny got leaner, faster, and feathered before taking to the air. Right now they are pecking seed from the feeder dangling outside my window. Simple, elegant—parsimonious, as paleontologists like to say—and yet from a nationalist view you can understand why it was ignored. Dinosaurs evolving into the larger family of life instead of going down in a blaze of intergalactic holocaust? No way.

But there was another side to this revival. At the time, America's status as superpower, as well as keeper of dinosaurs, was beginning to be challenged by China. Amid reports about China's theft of nuclear secrets, clandestine arms sales, and independent space exploration, word came that China was building the world's largest dinosaur park in Chuanjie province. Authorities there had discovered two skeletons of dinosaurs fossilized in mid-battle—an extremely rare find of obvious commercial appeal.

"Chuanjie has now passed Utah in the United States," Chinese papers continually boast (quoting an American expert) "to become the largest burial ground of dinosaurs in the world." A staggering announcement, given China's previous clumsy attempts to enter the paleontological major league. A 1983 China find, for instance, was named *Gongbusaurus*, literally, "Ministry of Public Works-osaurus."

And now a truly brilliant Chinese breakthrough—the discovery of a feathered dinosaur, and possible proof of Ostrom's theory—was visiting America. In the exhibit, these specimens were not mere bones but taxidermically dressed up, as if stuffed after a recent hunt, feathered from head to toe in harvest colors of sedge brown, crimson red, and dark yellow, all posed in the most aggressive postures possible: raised claws, open teeth, wings volant. They looked like enraged ranged turkeys. And hidden right there in the taxonomic name was their true significance: *Sinornithosaurus millenii*—"Chinese bird-reptile of the new millennium." Freshly discovered, freshly minted—a brand-new, slimmed-down dinosaur metaphor had sprouted wings and flown off to Asia. Maybe, it seemed, America was bowing out of the game.

But then maybe not. A few weeks after the exhibit arrived in Washington (Chinese dinosaurs in the bosom of our capital—the horror, the horror!), paleo-patriots could breathe easier. Scandal erupted when it was charged that one of the exhibits was a fake. Under headlines "Piltdown Bird" and "Buyer Beware," articles wallowed in new information that Chinese peasants were cobbling together different fossils, often with glue and paint, to feed the international market. The implication was clear. Like videocassettes in a free-trade dispute, Chinese fossils were just more cheap pirated fakes being dumped in the lucrative American marketplace.

Scarcely a few weeks later came an announcement from Hollywood, Florida. A philanthropist named Michael Feinberg had purchased a fossil for a museum there, and it had been closely reexamined. In the words of a breathless AP reporter, it was "a 75-million-year-old creature with a roadrunner's body, arms that resembled clawed wings and hair-like feathers." You want feathers? America's got your feathers right here. "A dinosaur Rosetta stone," said a museum associate, just in case anyone underestimated its significance. The Linnaean name of America's new proof of birdness was as rich in meaning as *Sinornithosaurus millenii*. In its own moist and Disneyesque way, the new find reincarnated that old blend of the bront's amiability with T. rex's dormant ferocity: *Bambiraptor*.

Our Mona Lisa, as Ostrom described it. A sentimentalized dinosaur for a sentimentalized time. In the tradition begun with *Diplodocus carnegii* a century ago, the full name of the latest celebrity dinosaur is *Bambiraptor feinbergi*.

Driving a stake through the heart of the Chinese bird-dinosaur has since become a seasonal blood sport. Last year, China announced a true birdlike dinosaur discovery. They called it *Protopteryx* (meaning "first feather"). As if, Professor Alan Feduccia of the University of North Carolina at Chapel Hill looked at the evidence and immediately laughed it off as "dino-fuzz."

But feathers hardly matter anymore because paleontologists are still cooing over the latest proof of dinosaur warm-bloodedness. This confirmation, found last year, derives from the discovery of the first intact dinosaur organ—a heart—that had miraculously survived fossilization. Described as a grapefruit-sized, reddish-brown stone, it was found in the heart of the heart of the country: South Dakota. True dino-land. America—where modern dinosaurs first stirred and where they still thrive. Reddish brown, the description reads, as if the blood were still in it, almost beating. That's how it goes with dinosaurs. We're always getting closer to the true dinosaur, the next dinosaur, the best dinosaur. It's America's task. There is always another one on the way. With only a fourth of the dinosaur fossils estimated to have been found, the empire has a ways to go.

Each subsequent discovery will conceal new messages in its bones, hints of our superpower's new place in this world and our hearts. The new bones will stand up and the old ones will lie down. The theories will wax and wane. But no matter what we may think the newest dinosaur means for that month, or that decade, it will really be about what every dinosaur has always been about—not extinction but the other, deeper dream of this nation: the big comeback, the perpetual *novus ordo* of America, the unexpected feat of resurrection.

FOOTNOTES

1. In Tim Hames's best-selling book, *Walking with Dinosaurs*, the author tries his hand at sustaining T.'s ebbing machismo by using a tactic of contemporary memoir publishing, the sexually explicit detail. In Haines's imagination, here are two T.s doing the nasty: "She raises her tail and he approaches quickly from behind. When he attempts to mount her, he uses his tiny forelimbs to steady himself by hooking into the thick hide of her back. The coupling is brief, but it is the first of many. . . . By staying close, he . . . increases his chances of fertilizing her by mating repeatedly. . . ." That's our boy.

2. Those two dinosaurs continue to reside at the American Museum of Natural History. In a 1995 redesign of the museum's expanded collection, the original duo were set aside in a separate room—the dinosaur of dinosaur exhibits. The original armature still holds up those bones, and that old blackened plumbing is easily as beautiful as the fossils themselves. The great dinosaurs had lain down 65 million years ago, and what put them once again on their feet? The tensile strength of Pittsburgh steel.

Extreme Dinosaurs[*]

By John Updike
National Geographic, December 2007

Before the 19th century, when dinosaur bones turned up they were taken as evidence of dragons, ogres, or giant victims of Noah's Flood.

After two centuries of paleontological harvest, the evidence seems stranger than any fable and continues to get stranger. Dozens of new species emerge each year; China and Argentina are hot spots lately for startling new finds. Contemplating the bizarre specimens recently come to light, one cannot but wonder what on earth Nature was thinking of. What advantage was conferred, say, by the ungainly eight-foot-long arms and huge triple claws of *Deinocheirus*? Or, speaking of arms, by *Mononykus*'s smug dependence on a single, stoutly clawed digit at the end of each minimal forearm? Guesses can be hazarded: The latter found a single stubby claw just the thing for probing after insects; the former stripped the leaves and bark from trees in awesome bulk. A carnivorous cousin, *Deinonychus*, about the size of a man, leaped on its prey, wrapped its long arms and three-fingered hands around it, and kicked it to the death with sickle-shaped toenails.

Tiny *Epidendrosaurus* boasted a hugely elongated third finger that served, presumably, a clinging, arboreal lifestyle, like that of today's aye-aye, a lemur that possesses the same curious trait. With the membrane they support, the elongated digits of bats and pterosaurs enable flight, and perhaps *Epidendrosaurus* was taking a skittery first step in that direction. But what do we make of such apparently inutile extremes of morphology as the elaborate skull frills of ceratopsians like *Styracosaurus* or the horizontally protruding front teeth of *Masiakasaurus knopfleri*, a late Cretaceous oddity recently uncovered in Madagascar by excavators who named the beast after Mark Knopfler, the lead singer of the group Dire Straits, their favorite music to dig by?

Masiakasaurus is an oddity, all right, its mouth bristling with those slightly hooked, forward-poking teeth; but, then, odd too are an elephant's trunk and tusks, and an elk's antler rack, and a peacock's tail. A difficulty with dinosaurs is that we can't see them in action and tame them, as it were, with visual (and auditory and olfactory) witness. How weird might a human body look to them? That thin and featherless skin, that dish-flat face, that flaccid erectitude, those feeble, clawless five digits at the end of each limb, that ghastly utter lack of a tail—ugh. Whatever did this creature do to earn its place in the sun, a well-armored, nicely specialized dino might ask.

Dinosaurs dominated the planet's land surface from some 200 million years ago until their abrupt disappearance, 135 million years later. The vast span of time boggles the human mind, which took its present, *Homo sapiens* form less than 200,000 years ago and began to leave written records and organize cities less than 10,000 years in the past. When the first dinosaurs—small, lightweight, bipedal, and carnivorous—appeared in the Triassic, the first of three periods in the Mesozoic geologic era, the Earth held one giant continent, Pangaea; during their Jurassic heyday Pangaea split into two parts, Laurasia and Gondwana; and by the late Cretaceous the continents had something like their present shapes, though all were reduced in size by the higher seas, and India was still an island heading for a Himalaya-producing crash with Asia. The world was becoming the one we know: The Andes and the Rockies were rising; flowering plants had appeared, and with them, bees. The Mesozoic climate, generally, was warmer than today's, and wetter, generating lush growths of ferns and cycads and forests of evergreens, ginkgoes, and tree ferns close to the Poles; plant-eating dinosaurs grew huge, and carnivorous predators kept pace. It was a planetary summertime, and the living was easy.

Not *that* easy: Throughout their long day on Earth, there was an intensification of boniness and spikiness, as if the struggle for survival became grimmer. And yet the defensive or attacking advantage of skull frills and back plates is not self-evident. The solid-domed skull of *Pachycephalosaurus*, the largest of the bone-headed dinosaurs, seems made for butting—but for butting what? The skull would do little good against a big predator like *Tyrannosaurus rex*, which had the whole rest of *Pachycephalosaurus*'s unprotected body to bite down on. Butting matches amid males of the same species were unlikely, since the bone, though ten inches thick, was not shock-absorbent. The skulls of some pachycephalosaurs, moreover, were flat and thin, and some tall and ridged—bad designs for contact sport. Maybe they were just used for discreet pushing. Or to make a daunting impression.

An even more impractical design shaped the skull of the pachycephalosaurid *Dracorex hogwartsia*—an intricate sunburst of spiky horns and knobs, without a dome. Only one such skull has been unearthed; it is on display, with the playful name derived from Harry Potter's school of witchcraft and wizardry, in Indianapolis's Children's Museum. Duck-billed *Parasaurolophus walkeri*, another late Cretaceous plant-eater, sported a spectacular pipelike structure, sweeping back from its skull, that was once theorized to act as a snorkel in swimming. But the tubular crest had no hole for gathering air. It may have served as a trumpeting noisemaker, for herd communication, or supported a bright flap of skin beguiling to a *Parasaurolo-*

phus of the opposite gender. Sexual success and herd acceptance perpetuate genes as much as combative prowess and food-gathering ability.

Dinosaurs have always presented adaptive puzzles. How did huge herbivores like *Brachiosaurus, Apatosaurus,* and *Diplodocus* get enough daily food into their tiny mouths to fill their cavernous guts? Of the two familiar dinosaurs whose life-and-death struggle was memorably animated in Walt Disney's 1940 *Fantasia* (though in fact they never met in the corridors of time, failing to overlap by fully 75 million years), *T. rex* had puzzlingly tiny arms and *Stegosaurus* carried on its back a double row of huge bony plates negligible as defensive armor and problematic as heat controls. Not that biological features need to be efficient to be carried along. Some Darwinian purists don't even like the word "adaptive," as carrying a taint of implied teleology, of purposeful self-improvement. All that is certain is that dinosaur skeletons demonstrate the viability, for a time, of certain dimensions and conformations. Yet even Darwin, on the last page of *The Origin of Species*, in summing up his theory as "Natural Selection, entailing Divergence of Character and the Extinction of less-improved forms," lets fall a shadow of value judgment with the "less-improved."

In what sense are living forms improvements over the dinosaurs? All life-forms, even such long-lasting ones as blue-green algae and horseshoe crabs and crocodiles, will eventually flunk some test posed by environmental conditions and meet extinction. One can safely say that no dinosaur was as intelligent as *Homo sapiens*, or even as chimpanzees. And none that are known, not even a heavyweight champion like *Argentinosaurus*, was as big as a blue whale. One can believe that none was as beautiful in swift motion as a cheetah or an antelope, or as impressive to our mammalian aesthetic sense as a tiger. But beyond this it is hard to talk of improvement, especially since for all its fine qualities *Homo sapiens* is befouling the environment like no fauna before it.

The dinosaurs in their long reign filled every niche several times over, and the smallest of them—the little light-boned theropods scuttling for their lives underfoot—grew feathers and became birds, still singing and dipping all around us. It is an amazing end to an amazing evolutionary story—*Deinonychus* into dove. Other surprises certainly lurk within the still unfolding saga of the dinosaurs. In Inner Mongolia, so recently that the bones were revealed to the world just this past spring, a giant birdlike dinosaur, *Gigantoraptor*, has been discovered. It clearly belongs among the oviraptorosaurs of the late Cretaceous—90-pound weaklings with toothless beaks—but weighed in at one-and-a-half tons and could have peered into a second-story window. While many of its fellow theropods—for example, six-foot, large-eyed, big-brained *Troodon*—were evolving toward nimbleness and intelligence, *Gigantoraptor* opted for brute size. But what did it eat, with its enormous toothless beak? Did its claw-tipped arms bear feathers, as did those of smaller oviraptorosaurs?

The new specimens that emerge as tangles of bones embedded in sedimentary rock are island peaks of a submerged continent where evolutionary currents surged back and forth. Our telescoped perspective gives an impression of a violent struggle as anatomical ploys, some of them seemingly grotesque, were desperately tried and eventually discarded. The dinosaurs as a group saw myriad extinctions, and the final extinction, at the end of the Mesozoic, looks to have been the work of an asteroid. They continue to live in the awareness of their human successors on the throne of earthly dominance. They fascinate children as well as paleontologists. My second son, I well remember, collected the plastic dinosaur miniatures that came in cereal boxes, and communed with them in his room. He loved them—their amiable grotesquerie, their guileless enormity, their unassuming small brains. They were eventual losers, in a game of survival our own species is still playing, but new varieties keep emerging from the rocks underfoot to amuse and amaze us.

CARNOTAURUS

X-FACTOR Bull horns, tiny arms
WHEN 82–67 million years ago
WHERE Argentina

Consider the evolutionary hand dealt to *Carnotaurus*, or "meat-eating bull": a big, bad, but seemingly underequipped predator, as if nature had set out to design a perfect killing machine but ran out of funding. Powerful jaws and long, agile legs suggest a highly mobile hunter prowling the lake-shores of what is now Patagonia.

Its skull, constructed like a battering ram, features a stout pair of horns. Yet accompanying this formidable hardware are tiny arms (even more stunted than the famously puny arms of *Tyrannosaurus rex*) and surprisingly small teeth. Some scientists, like University of Chicago paleontologist Paul Sereno, envision *Carnotaurus* and its kin as dinosaurian hyenas—fleet of foot and short-snouted to track down and gnaw on carcasses. "Who needs arms for that?" he asks.

PARASAUROLOPHUS

X-FACTOR Trombone crest
WHEN 76 million years ago
WHERE North America

The tubular bone sweeping back from a *Parasaurolophus walkeri* skull has inspired a variety of theories about its function. Weapon? Breathing tube? Hypersensitive nose? None of the above? Aided by computer modeling, scientists now think it was used to generate sounds like a trombone, though it also may have played a role in sexual display.

X-FACTOR Inscrutable teeth
WHEN 70-65 million years ago
WHERE Madagascar

The mouth of *Masiakasaurus* speaks to how this German-shepherd-size meat-eater survived in the river basins of northwestern Madagascar, near the end of the dinosaurs' reign. But what is it saying? Stony Brook University paleontologist David Krause led the team that found the remains, including part of the lower jaw. *Masiakasaurus* has long, conical front teeth with hooked tips that curl out of its mouth—unique among theropods—while its back teeth are more typically blade-like and serrated. So how did it use such a specialized mouth? "Our best guess is the teeth up front were used to stab small prey, perhaps mammals, lizards, and/or birds," says team member Scott Sampson of the University of Utah, "and the teeth at the rear of the jaw were then used to tear up the kill." Despite its formidable dentition, *Masiakasaurus* was likely prey itself for crocs and other large carnivores, like the 20-foot-long theropod *Majungasaurus*, with which it shared territory. Against such monsters, its best defenses would have been speed and agility.

X-FACTOR High-spined giant
WHEN 97 million years ago
WHERE North Africa

In 1912 a collector for German paleontologist Ernst Stromer emerged from the Egyptian desert with the remains of the biggest predatory dinosaur the world had ever seen. The creature may have measured more than 50 feet long—arguably still the largest terrestrial carnivore known. It possessed crocodilelike teeth and a row of enormous spines (some six feet long) projecting from the vertebrae, which prompted Stromer to name the beast *Spinosaurus*. Scientists have argued over the function of the spines ever since. The debate offers insight into one of the key questions paleontologists face: How do you reconstruct a flesh-and-blood dinosaur from just a few bones?

One way is to piece together clues by comparing the new specimen with more complete skeletons already in hand. Scientists also draw inferences from the extinct animal's environment and the way living creatures function with analogous skeletal equipment. Naturally, the less one has of a specimen, the more speculative become the reconstructions—and the more heated the controversy surrounding them. Over the years, many scientists have argued that *Spinosaurus*'s vertebral projections were connected by a fleshy membrane. Some living lizards employ similar "sails" for sexual display. Perhaps *Spinosaurus* too sported a sail to win the attention of mates, as some paleontologists hypothesize about *Amargasaurus*. The sail may also have helped *Spinosaurus* regulate body temperature, serving as a radiator to cool the blood, much as a car radiator cools the water circulating through an engine.

Then again, perhaps the various renderings of a sail-bearing *Spinosaurus* have all missed the mark. As Smithsonian paleontologist Hans-Dieter Sues points out, other related dinosaurs, such as *Baryonyx*, regulated their body temperature just fine, sans sail. Sues seconds a notion put forth a decade ago by Jack Bowman Bailey of Western Illinois University: The spines instead supported a structure similar to a bison's hump. "If *Spinosaurus* had puny, slender spines, they might have supported a sail, but these were very massive," says Sues. "It makes more sense that *Spinosaurus*'s spines were embedded in a lot of muscle and tissue."

Other scientists argue that humps are usually found on herbivores in arid environments, while *Spinosaurus*, a carnivore, appears to have been living in a coastal mangrove forest. Paleontologist Paul Sereno ascribes to the theory that the spines supported a sail for sexual display. *Suchomimus*, a closely related predecessor, had much smaller vertebral spines. Millions of years later, says Sereno, "*Spinosaurus* took that trait to the extreme."

TUOJIANGOSAURUS

X-FACTOR Shoulder spikes
WHEN 161–155 million years ago
WHERE China

With a thorny tail and rows of bony plates along its back, *Tuojiangosaurus*, like its better known cousin *Stegosaurus*, resembles a Jurassic tank. What grants this ponderous Chinese herbivore admission to the ranks of the truly bizarre, however, is the long, tapering spike thrusting out from each shoulder. "The shoulder spikes would have helped protect its vulnerable flanks, which would have been right at the level of an attacking allosaur," says University of Maryland paleontologist Thomas Holtz. As for the plates on its back, their purpose is a matter of much debate, says Susannah Maidment, a paleontologist at Cambridge University. Early armored dinosaurs were covered with small scutes to protect against a predator's bite, a trait passed on more or less unchanged to some of their descendants. But in others such as *Tuojiangosaurus*, the scutes gave way to plates along the backbone, which perhaps made the animal look bigger, but offered little protection. A large theropod, says Maidment, would have been able to chomp through them "like potato chips."

NIGERSAURUS

X-FACTOR Shovel-like mouth, 600 teeth
WHEN 110 million years ago
WHERE North Africa

Does natural selection ever paint a species into a corner, leaving it with anatomy so specialized that a slight change in its environment pushes it over the edge into extinction? Consider the 50-foot-long diplodocoid (a branch of the sauropod group) *Nigersaurus*—an anatomical sideshow with a mouth shaped like a vacuum cleaner, hundreds of tiny teeth, a boom of a neck, and skull bones thin to the point of translucence. How did it survive with such a preposterous eating apparatus? Paleontologist Paul Sereno, a National Geographic explorer-in-residence, picked up the quizzical beast's trail in the mid-1990s in northeastern Niger. *Nigersaurus*'s oddest feature is its broad, straight-edged muzzle, which allowed the business end of its mouth to work very close to the ground, "like a lawn mower," says Sereno. Its 600 teeth, each about the size of a toddler's incisor, were tightly aligned, with a single row of more than 50 in operation in each jaw at any one time. A CT scan exposed up to eight "replacements" stacked up behind each tooth, so that new teeth immediately filled in for worn ones. Despite its impressive battery of teeth, *Nigersaurus* had a weak bite. Where the jaw muscles attach to the skull, the bone is as thin as a paper plate.

"Its mouth appears designed for nipping rather than chomping or chewing," says Sereno, pointing to wear patterns that suggest the teeth slid by one another like a pair of shears. With nearly 80 percent of the animal's skeleton recovered, a portrait emerges of a finely tuned eating machine designed to crop mouthfuls of soft plants growing near the rivers that coursed through what is now the Sahara's southern flank. Its long neck would have allowed *Nigersaurus* to mow down an entire field of plants without taking a single step.

A number of these features, seen in the extreme in *Nigersaurus*, show up over millions of years in other diplodocoids, which thrived on nearly every major landmass. That, says Sereno, suggests that this feeding strategy emerged in primitive form much earlier. But could this evolutionary trend toward the perfect eating machine ultimately have led to the extinction of the lineage? "Selection can favor specialization that can be advantageous over the short run, but create vulnerabilities over the long haul," says University of Chicago evolutionary biologist David Jablonski. We'll never know for sure whether *Nigersaurus* fell victim to its own outlandishness. But while it lasted, this fern-eating giant was a bizarre and beautiful success.

STYARACOSAURUS

X-FACTOR Massive horned frill
WHEN 75 million years ago
WHERE North America

Like an armor-laden knight, *Styracosaurus* would have cut an imposing figure on the forested river plains in what is now Alberta, Canada. Multiple individuals of these rhino-size herbivores have been identified in the same bone beds, suggesting they traveled in herds. Horned dinosaurs are a well-understood group, says Hans-

Dieter Sues, and since *Styracosaurus* lived near the end of this lineage, we can trace the evolutionary paths that led to it. "Its ancestors began with a little bump over their nose and then developed a little bit of a frill at the back of the skull," says Sues, "but *Styracosaurus* takes these traits to the top." The bump on the nose in ancestral species evolved into an enormous spike that would have given *Styracosaurus* a potent weapon to fight off predators and fend off rivals. Meanwhile, the skull frill enlarged and added a profusion of horns, which probably let other styracosaurs identify it from a distance. Some scientists have suggested blood pumped into the skin covering the frill could have caused it to change color, possibly to attract mates or to scare enemies. "These extreme traits just didn't suddenly appear," says Sues. "There were compelling reasons why they were selected and pushed down the evolutionary line."

Flesh & Bone[*]

A New Generation of Scientists Brings Dinosaurs Back to Life

By Joel Achenbach
National Geographic, March 2003

The bull gator lay in the sand under the oak trees. A few days earlier he had been hauled out of a murky lake in central Florida. Researchers instantly named him Mr. Big. He was sofa size, with fat jowls framing his head like a couple of throw pillows. He would have measured thirteen and a half feet if a rival hadn't chomped the last foot off his tail.

Four people sat on his back. Excited alligators do more than thrash—they can spin like wound-up rubber bands. Yet Mr. Big, with his mouth taped shut and a towel over his eyes, was completely docile, as inert as luggage. He behaved like a gator basking in the sun rather than one in the middle of a science experiment.

Gregory Erickson, the scientist, stood a few paces away, grimly holding a plastic pole tipped with a little square plate called a force transducer. He intended to put this in the animal's mouth, to measure the force of its bite. Erickson also intended to retain all his body parts, which explained his serious countenance.

A man on the gator's back removed the towel and the tape. The animal opened its eyes and hissed. It was a factory noise, a steam pipe venting. The mouth opened as slowly as a drawbridge. The maw on Mr. Big was spacious enough to house a poodle. Erickson presented the force transducer to the largest tooth at the back of the right upper jaw—and the jaws snapped shut.

"Trouble, we got trouble!" Erickson said as the gator, pole firmly clamped, began to lurch in his direction. But then the animal calmed down. Erickson read some numbers off a meter. "Two point nine six—that's a lot!" he said.

The creature's jaws had come together with nearly 3,000 pounds of force.

The odd thing about this little experiment was that it was fundamentally about dinosaurs. Erickson, a paleobiologist at Florida State University, is an expert in the feeding behavior of tyrannosaurs, including the bite marks left on bones. That research spurred him to find out more about bites in general, which is why he's out here moonlighting with crocodilians.

We loaded Mr. Big on an airboat to be towed overland back home to Lake Griffin, about an hour to the south. We spent the night lakeside, testing ten more gators that had been freshly yanked from the water. All the while, we talked about dinosaurs.

Shortly before dawn—by which time we were thoroughly scuffed up and swampy, though still in possession of all our digits—Erickson turned to me and said, "It's not like diggin' bones, is it?"

Bone-diggin' is still essential, but an increasing number of vertebrate paleontologists are going beyond the bones, looking for novel ways to study dinosaurs.

Instead of spending the summer in a dusty badlands bone bed, they might spend it in a laboratory, analyzing the evolution of the flight stroke by tossing pigeons into a wind tunnel. Instead of scraping away the sandstone overburden on a nicely articulated ceratopsian, they might point and click on a computer screen, pivoting digital bones.

These paleontologists tend to be on the young and idealistic side, determined to intensify the scientific rigor of their profession. Their goal is to hunt not just for dinosaurs but for something even harder to reconstruct—how dinosaurs functioned and behaved.

They are tackling difficult questions:

Were dinosaurs fleet of foot or ponderous?

What did they eat?

Did they hunt or migrate in packs?

Did they parent their young?

How fast did they grow? Did they get bigger and bigger even into old age? And how old did they get?

Did they use horns and frills and spikes in battle, like they do in the movies? Were these unusual anatomical structures part of the business of attracting mates?

How did one group of these creatures develop the ability to fly?

These new scientists are a diverse bunch, emerging from evolutionary biology, biomechanics, botany, physiology. Their tools include computers, CT scans, x-rays, and electron microscopes. They publish papers with titles like "Nostril Position in Dinosaurs and Other Vertebrates and Its Significance for Nasal Function" and "Caudofemoral Musculature and the Evolution of Theropod Locomotion." We might say they are geekier than the older generation of dinosaur researchers, and then quickly add that we mean this in the best sense of the word.

Make no mistake, the "field"—which is anywhere and everywhere bones can be found—still dominates dinosaur research. In recent years the field has produced feathered dinosaurs from China, egg-laying dinosaurs from Patagonia, and a host of new dinosaur species, such as the scale-breaking *Argentinosaurus* and the fearsome *Giganotosaurus*. In the field we find direct evidence of a lost world ruled by titanic creatures, thriving all over the globe from 230 to 65 million years ago, during the Mesozoic era.

And the field has charms that the laboratory can't match. The field is the staging ground for that whole Indiana Jones thing, for the type of charismatic, rock-star scientists who hang out in dinosaur graveyards with shovels, picks, plaster, graduate students, and personal documentary film crews.

But perhaps the very glamour of dinosaurs has spawned the backlash, the willful retreat to scientific basics by Greg Erickson and researchers like him. Most scientific disciplines aren't caught in the gravity well of public fascination. If you study fossil mollusks, for example, you aren't likely to be asked to become a scientific adviser for a Hollywood blockbuster. No one has snail fever. But dinosaur fans are insatiable for information. The new generation of scientists wants to put constraints on all the hypotheses flying around, and they think that the truth about dinosaurs—and dinosaur behavior—won't be uncovered with bones alone.

"For 20 years we've done what we call arm-waving," says Jack Horner, a legendary bone collector. "We've made hypotheses based on very little evidence. Now we're sitting down, we're saying, 'We've got all these ideas, are they real?'"

Horner can arm-wave like a champ, as he will admit. Since 1991 he's been arguing, for example, that *Tyrannosaurus rex*, the very emblem of predation, the killer of killers, was actually just a scavenger, an eater of the dead. An overgrown turkey vulture! Those banana-size teeth weren't for ripping live flesh, says Horner, they were for crushing the bones of a carcass. This is vintage contrarianism, and Horner so far has failed to persuade many of his peers, who point out that *T. rex* need not have been one thing or another. Hyenas, for example, are scavengers one day, predators the next.

But in any case this is precisely the kind of argument that can't be won by speaking louder than one's opponents. Science requires data. Science requires that ideas be subjected to tests. And paleontology—if the new generation has its way—will be seen as a no-nonsense field, a hard science, in addition to being a thrilling subject built around the bones of large, scary animals.

"This is where we have the rhino heads and the manatee heads," Lawrence Witmer is saying. "We've got a whole bin of ostrich heads and necks. We've got ducks and geese. Here's a bag of alligator parts."

We're in the deep freeze of his laboratory at Ohio University in Athens. Witmer has quite the collection of heads. They belonged to creatures that died, or were killed for some other reason, and were then obtained by Witmer for research. I keep thinking Witmer is about to produce the horse head from *The Godfather*.

Witmer reconstructs the soft tissues in dinosaur heads. His method exploits similarities among creatures across the vastness of time. It turns out that a dinosaur of the Triassic period, 248 to 206 million years ago, had anatomical features remarkably similar to those of a contemporary alligator or seagull.

Witmer recently caused a stir when he said that artists had long put the nostrils of dinosaurs too high on the head. He spent months studying the relative positions of noses and nostrils in modern animals. He wanted to see if there is a correlation—whether the bone of the nose reveals the location of the fleshy nostril. He found that as a rule there is such a correlation.

Witmer then examined fossils and discovered that in modern renditions of dinosaurs the nostrils had always been misplaced. They should be shown low on the nose, near the mouth. Nostrils in that location would heighten the animal's ability to nuzzle a potential food item and decide whether it was biteworthy.

Witmer investigated another paleontological presumption, the notion that *Triceratops* and other plant-eating dinosaurs had cheeks, like cows or horses or humans. Conventional wisdom said these cheeks were like feed bags, helping the animal chew and re-chew vegetation. Witmer, to his surprise, discovered that animals with cheeks have bone structures that are lacking in *Triceratops* and other herbivorous dinosaurs. *Triceratops*, he thinks, had something more like a bill or a beak.

The plant-eating dinosaurs may have clipped vegetation off plants with these beaks and then swallowed the material pretty much intact. "They probably actually chewed with their stomachs," Witmer says.

The day I visited, Witmer took maybe 15 animal heads out of the freezer and arranged them on a table, a buffet from a nightmare. He explained how he dissects them to examine soft tissues and how he uses his findings to flesh out model dinosaur heads. As we talked, the heads thawed. They got rather . . . drippy. Beyond ripe. The moose head seemed particularly malodorous. "Most of these guys are past their sell-by date," Witmer said, unfazed.

A few hours later, putting the heads back in the freezer and mopping up the mess, he said, "There's no real substitute for doing what we're doing—getting your hands dirty, rolling up your sleeves, getting out a scalpel, and seeing how these things are really put together."

Stephen Gatesy is another pioneer of the new dinosaur science and can spend days at his computer screen zeroed in on a single trochanter, the knobby protrusion on a bone where a tendon once attached. He might spend months or even years on a shoulder joint.

"I'm not ambitious enough to take on the whole animal," he told me when I visited him at Brown University.

That's a classic statement of the new science. Think of how dinosaur paleontology has been dominated by "the whole animal," by spectacular specimens, huge skeletons that can fill the entrance hall of a museum. This fellow Gatesy can get wrapped up in a single metatarsal.

A traditional dinosaur researcher might take a couple of loose dinosaur bones, stick them together at the joint, wiggle them, pivot them, move them around, and pronounce, "I think they went like this." Gatesy wants to do the hard labor of figuring out how these structures evolved and affected locomotion—how dinosaur ancestors, for example, went from walking on four legs to walking on two (and apparently back to walking on four in some cases). Thrown into the mix is the stunning fact that some dinosaurs lifted off the ground entirely.

How did flight develop?

Did the first airborne dinosaurs merely glide, or did they flap?

Did the flight stroke evolve from other types of motion, such as grabbing prey or trying to elude a predator? Were they flappers before they were fliers?

Or did flying emerge from climbing? The flight stroke might have given an animal increased traction on steep inclines. There's recent research at the University of Montana showing that baby birds, for example, start flapping when ascending an incline. They're not trying to fly, just trying to climb better.

Anatomy can be deceiving. Birds have hollow bones, feathers, wings, reduced tails, and wishbones, each characteristic designed for flight. And yet each of these traits or something like it appears in the fossil record before birds flew.

The dinosaur fossil record is actually rather poor. Intact, articulated, museum-quality skeletons are fairly rare. Fossils fall apart. A bone exposed to the elements may simply explode. In some bone beds there are so many tiny skeletal fragments you'd think the creatures had been dropped from a plane.

That's why dinosaur behavior is so difficult to fathom from just bones—why the task of understanding dinosaurs is truly like trying to squeeze blood from a stone. Some would argue that dinosaur behavior is a topic all too similar to extraterrestrial life—long on speculation and short on data.

Gatesy and other paleobiologists are struggling to ascend what Witmer calls the Inverted Pyramid of Inference. Imagine an upside-down pyramid with, at the pointed bottom, the word "bones." Bones are the known commodity, the solid evidence. They are aged; they may be broken, cracked, ambiguous. But you can at least hold them in your hand.

Above bones on the inverted pyramid are soft tissues. There aren't many of those because they rarely fossilize.

Above that is function, how the bones and tissues worked.

Above that—so very far from the hard evidence of bones—is behavior.

Above that is environmental interaction. The dream would be to know the behaviors of many different dinosaurs and to be able to put them in context so you'd know what dinosaurs ate and where they slept and what they feared and how they prowled the landscape.

And at the very top of the inverted pyramid, as far from hard science as you can get, is . . . well, probably the purple dinosaur known as Barney.

Dinosaur science was inherently flamboyant and mind-boggling from its very beginning. In 1853 paleontologist Richard Owen (who had given dinosaurs their name a little more than a decade earlier) staged a celebrated sit-down dinner in London. He and 21 other scientists dined at a table set up inside a model of an *Iguanodon*. An engraving of the scene created a national sensation in Great Britain.

Bone hunters scrambled to find ever more spectacular specimens. By the early 20th century the preeminent ambition in the field was to mount a skeleton dramatic enough to scare the bejabbers out of a schoolchild.

Roy Chapman Andrews's journeys to Mongolia's Gobi in the 1920s were worthy of a Cecil B. DeMille movie—great caravans of camels stretching into the wasteland, with Andrews packing a pistol and posing on bluffs with jaw thrust forward.

But just two decades later the heroic age of dinosaur collecting was over. Scientists began to view dinosaurs as an evolutionary dead end. They were tail-dragging losers in a Darwinian world, defeated by quicker, smarter mammals. "It is a tale of the triumph of brawn, a triumph that was long-lived, but which in the end gave way to the triumph of brain," wrote Edwin H. Colbert of the American Museum of Natural History in his 1945 publication, *The Dinosaur Book*.

The field was in the doldrums when in 1975 Robert Bakker, a maverick paleontologist at Harvard, published an article in *Scientific American* titled "Dinosaur Renaissance." It gave a shot of adrenaline to the entire discipline. Bakker, building on the work of his mentor, Yale's John Ostrom, said that dinosaurs were not cold-blooded, but rather warm-blooded, active, quick. They may have nurtured their young and hunted in packs. And they weren't even extinct! Birds, Bakker said (again, echoing Ostrom), are themselves the direct descendants of dinosaurs.

The new image carried the day. In the *Jurassic Park* movies the dinosaurs are fully Bakkerian. They sprint across meadows. They nurture their young. The "raptors" are so savvy they seem on the verge of inventing spaceflight.

"I'm a method paleontologist," Bakker said one day in Boulder, Colorado. He means like a method actor—like Robert De Niro or Marlon Brando. "I want to be Jurassic. I want to smell what the megalosaur smells, I want to see what he sees."

From a cigar box on the table he pulled a *T. rex* tooth. "This is a bullet," he said. A fossil site is a crime scene, he explained, and the teeth are the bullets. He thinks he's found evidence in the teeth of allosaurs, meat-eaters of the Jurassic, that those dinosaurs dragged huge prey to their nests to feed to their babies. As he puts it, "You're under care. Your first meal is given to you, and it's steak, and it's six feet thick."

Since Bakker's been theorizing about dinosaurs for more than a quarter century, I asked him if he thought dinosaur paleontology has become a more rigorous science with its new emphasis on lab work. He replied by citing an 1822 study of Ice Age hyenas by the Reverend William Buckland. Bakker says Buckland's work is as good as any modern paleontology. The profession looks at history, Bakker says—at the narrative of the rocks and bones. It can't possibly turn into a laboratory science.

"People who don't understand paleontology try to make it physics," he said. "Paleontology is history. It is made up of millions and millions of crimes. There are victims and there are perps."

Meanwhile, back at the lab. . . .

John Hutchinson, a 30-year-old researcher at Stanford, wants to answer a big question: Could *T. rex* run? If so, how fast? Did it have the leg muscles to sprint 45 miles an hour as some paleontologists contend?

Hutchinson doesn't think *T. rex* was that swift. He ponders *T. rex* through the prism of biomechanics. "The dream goal is to reconstruct exactly how an extinct dinosaur moved," he says. He uses a computer program that has digitized a number of *T. rex* bones.

To run that fast, Hutchinson figures, *T. rex* would have to have been almost all leg. Chickens run well, but a 13,000-pound chicken, Hutchinson has calculated, would need to have 62 percent of its mass in *each leg*.

Hutchinson also studies elephants and has made several trips to Thailand to analyze their locomotion. He paints white dots on elephants at crucial joints in the shoulders and legs. Then he chases the elephants, shouting "*Bai, Bai!*" which means "Go, Go!" In videotapes he captures the movement of the white dots.

What Hutchinson sees in the tapes doesn't look like running. At least not exactly.

"The best definition of walking is that the body swings over the leg like a stiff, inverted pendulum. Running is very different. It's like a pogo stick, a bouncing ball. Instead of the leg being stiff, it compresses like a spring. So you're using that spring to keep running efficiently. The spring stores energy."

There are a couple of intermediate forms of locomotion, including what has been called the Groucho run, named after the bent-legged walking that Groucho Marx made famous.

Elephants are more on the Groucho side of things. They keep at least one leg on the ground at all times—like a walker—but the white dots move down then up, indicating a bouncing gait.

Hutchinson showed me how to use his computer program to move muscles around, to attach them at different places on the bones, altering the leverage. By playing around, I'm pretty sure I created a dinosaur that couldn't do much of anything but fall over backwards.

"I could spend all my life on this, analyzing every little data point," he said. "You have to have very well-defined questions, or else you'll just get submerged in data and never get anywhere."

There's another way to observe dinosaur behavior, and it doesn't involve bones or computers. Dinosaurs left tracks.

One summer day I checked out some dinosaur tracks at an abandoned mine in the Rocky Mountain foothills near Grand Cache, Alberta, Canada. I was with Rich McCrea, a 33-year-old doctoral candidate at the University of Alberta, who has scrambled over almost every square inch of exposed rock.

McCrea likes tracks. He says they're the closest he can come to live dinosaurs without using a time machine. When he presented his master's thesis, he jokingly told the professors he'd love to find a six-footed trackway. You know, two dinosaurs mating. Reptilian passion captured in stone. The professors were not visibly amused.

The fact is, most dinosaur footprints capture a mundane activity: walking. In one direction, usually. One of McCrea's associates jokes that, to judge by most dinosaur tracks, these creatures couldn't turn.

The mine is at the end of a dusty road that until recently was heavily used by coal trucks. The miners sliced off a chunk of a mountain, and there's a wall of stone more than two miles long. At first you might not notice the prints. Then you see one or two, clearly outlined on the rock face. Then they gradually come into focus.

The rock face is covered with footprints—"totally polluted," McCrea says admiringly.

Some go this way, some go that way. There are quadrupeds and bipeds, plant-eaters and meat-eaters. Some tracks strongly suggest herding behavior, and McCrea thinks there's evidence that meat-eaters stayed clear of the deep muck of coal-producing marshes. On some of the dark rock surfaces, remnants of swampy terrain, there are only plant-eater tracks. They were probably ankylosaurs, McCrea says. "They're like Humvees, four-wheel drives."

We climbed up a seam of broken rock, feet churning through coal fragments, and on a higher rock face found some theropod tracks, footprints that quite possibly had never been seen by a human being. The mine, after all, was a fairly recent operation, and the rock faces sheer off regularly, meaning there are always new exposures. Yet after a few hours of exploring, one also sees the limits of the trackway profession. Footprints are just footprints. Other scientists have eventually given up on this site, McCrea says. "They couldn't handle the ambiguities inherent in footprint research."

A while back Steve Gatesy decoded some dinosaur footprints. He'd gone with some colleagues to Greenland—yes, even the most dedicated digital bone manipulators spend time in the field—and found thousands of tracks. They varied greatly, as though made by different species. Some tracks had an odd, bulbous structure at the end of the third digit, like a miniature volcano had erupted. What kind of feet left such odd prints?

For help, Gatesy turned to a turkey. He and a student at Brown bought one from a nearby farm and coaxed it to walk across a variety of hard and soft surfaces, including thick mud. The turkey didn't much care for this, but the tracks it created offered a revelation: All those different dinosaur footprints could have been made by the same species. What varied was not the type of animal but the type of surface.

And that odd, volcanic shape at the end of the third digit? The turkey and the mud explained that too. As the foot goes into the goop with toes spread, it makes the initial footprint. It strikes the hard subsurface then lifts again, bunching like a closed fist. The entire foot emerges from the muck at the front of the track, creating a craterlike exit mark.

It's arcane, to be sure, but science is often nothing more nor less than deconstructing what we're staring at.

The ultimate dinosaur behavior was the act of going extinct. And the mystery of that event has hardly been solved.

There's the easy, unsatisfactory answer: An extraterrestrial impact wiped them out. Quick, brutal, efficient. In documentaries there is the obligatory scene of the great impact, a flash of light, a blast wave, and the dinosaurs, blowing away like leaves before a storm.

But that can't be the whole story.

Philip Currie, his mind 75 million years in the past, roams the valley of the Red Deer River in Dinosaur Provincial Park in southern Alberta. He's the quintessential field scientist, and he's been hauling dinosaurs out of the badlands here for more than a quarter century.

I spent a day with Currie, stumbling over loose sandstones and mudstones, while he somehow remained perfectly upright, as though equipped with gyroscopes. It's not an easy workplace. Once a lightning bolt hit 30 feet from Currie and left a smoking crater in the sand. Rain turns the surface to grease. A person can get lost in about three minutes. One of his top technicians bounced out of the back of a pickup—a long story involving a misadventure at a saloon—and wandered all night while his colleagues looked for him in ditches along roadways. To treat himself to a fancy dinner, Currie goes to a bar in the town of Patricia—a community that looks like it narrowly escaped extinction itself.

Summers in this valley bring a bumper crop of knowledge. In one bone bed Currie found the skeletons of scores of centrosaurs. From the way the bones lay, it was clear to Currie that these creatures had died in water, and he inferred that they'd panicked while fording a swollen river. "This is probably the bone bed that got people talking about herding in a serious way," he said.

Currie has so far found 35 different species in Dinosaur Provincial Park. Farther up the Red Deer River, at Drumheller, where the fossils are around 70 million years old, there have been only about 20 species of dinosaurs found. And farther up the river, where the rock is 65 million years old, the last years of the Cretaceous period show only a few types of dinosaurs, including *T. rex*, *Triceratops*, and *Ankylosaurus*.

So even before dinosaurs became extinct, they were disappearing in this part of the world. That is why it's so important for the discipline to go beyond the bones and truly understand these creatures and their environment. Something triggered a tremendous decline in biodiversity. The big impact may have been merely the final blow.

The end of the Cretaceous was a time when the global climate was changing and the sea level dropping. A shallow sea that covered the heart of North America drained. Lands that were formerly separated by water were now connected. New species arrived, perhaps carrying deadly microbes.

No wonder it's such a haunting scenario. Our world today is undergoing a climate change, a period of emerging pathogens, a rapid mixing of Earth's biota, a loss of biodiversity, and a virtual shrinking of the entire planet.

Currie and I passed the afternoon in a remote part of the park, looking for new bone beds. We came upon a hillside covered with fragments, including some preserved inside unusually large nodules of ironstone. "Weird and different," Currie declared as he took a satellite reading of our position. Bone bed 185, he named it.

It might yield some answers. Or it might yield nothing but shoulder bones—the kind you look at for a second, then toss over your shoulder. But it was still exciting, because what we don't know about dinosaurs is far more than what we know. No matter how you practice it—with shovels or computer programs, with fossils or rhino heads from a freezer—this is still a new and evolving science. We've just scratched the surface.

Dinosaurs Under the Knife[*]

By Erik Stokstad
Science, November 5, 2004

With high-domed skulls built like battering rams, dinosaurs called pachycephalosaurs look for all the world as if they must have butted heads. Paleontologists imagined the males sparring for mates as bighorn sheep do, and the idea was bolstered by radiating bony structures that apparently strengthened the head against impacts. But did they actually knock noggins?

To find out, Mark Goodwin of the University of California, Berkeley, and John Horner of Montana State University in Bozeman did something that would give most museum curators the heebie-jeebies: They sawed open the skulls to examine the fossilized bone tissue. The answer was trapped within the domes, Goodwin says, and histology—the study of tissues—was the only way to get it.

Preserved in the bone, as in many fossils, was a beautiful record of the original tissue, down to the level of individual cells. That's beyond the resolution of computed tomography scanners. By studying pachycephalosaurs of various ages, Homer and Goodwin determined that the radial structures were ephemeral features associated with fast-growing bones of juveniles. There was no sign of stress to the skull bones, they reported in the spring issue of *Paleobiology*. "I didn't see any evidence that they headbutted," Goodwin says. However, he and Horner did find bundles of so-called Sharpey's fibers, which anchor ligaments and also thick pads of keratin to bone. Horner and Goodwin speculate that this may have secured a crest to the top of the head, perhaps for display.

More and more paleontologists are putting their fossils under the knife—the rock saw, actually—to gain new insights into their biology. "The microstructure includes a tremendous amount of information," says Armand de Ricqlès of the Université Paris VII. After removing a slice of bone, they glue it to a glass slide and then grind it until it is transparent. Studying this "thin section" of bone tissue with microscopes can explain the origin of strange structures, such as the thick heads of pachycephalosaurs and the plates of stegosaurs, and help test hypotheses about their function. "I get quite excited about the potential of using bone micro-structure to flesh ancient animals out and make them more real," says Anusuya Chinsamy-Turan of the University of Cape Town, South Africa.

Paleohistology is already shedding light on the question of how sauropods and tyrannosaurs attained their gigantic sizes and other evolutionary patterns. It can tell adult animals from juveniles, and it provides the only way to determine how old extinct animals were when they died and how quickly they grew—key questions for studying their population biology and ecology. "We're on the cusp of being able to learn a lot about the biology of these animals—things we thought we'd never be able to tease out of the bones," says Lawrence Witmer of Ohio University College of Osteopathic Medicine in Athens. This week at the Society of Vertebrate Paleontology (SVP) annual meeting in Denver, Colorado, paleontologists unveiled a bumper crop of histological studies, from a possible determination of the sex of a *Tyrannosaurus rex* to identification of an island of dwarf sauropods. "This field is about to explode," says paleobiologist Gregory Erickson of Florida State University, Tallahassee.

DIVERSE TISSUES

Paleohistology has a long history. For 150 years, paleontologists have used the technique to classify ancient fishes. But not until the early 20th century did they begin to compare the microstructure of various fossil land animals. Studies by Rodolfo Amprino of the University of Turin, Italy, led to an observation in 1947 that is now called "Amprino's rule": The rate of an animal's growth strongly influences the type of tissue deposited in its bones. A rapidly growing bone has many blood vessels. Its characteristic "fibrolamellar" texture is marked by quickly deposited fibers and holes that are then filled by bony structures called primary osteons. In contrast, a bone growing more slowly has a texture called lamellar-zonal with fewer blood vessels and a finely layered appearance.

Dinosaurs typically have bone tissue that more closely resembles the quickly growing bones of large birds and mammals than the slow, lamellar-zonal tissue of reptiles. In 1969 and in later papers, de Ricqlès suggested that the dinosaur bone tissue might indicate fast, continuous growth and an active metabolism like that of mammals and birds. This idea played an important role in the "renaissance" that changed the perception of dinosaurs from sluggish reptiles to active, possibly warm-blooded animals. Tissue type turned out not to be a simple indicator of metabolism, but it does indicate the general pace of growth.

Bone tissue offers another way to understand the growth of ancient animals. Bones sometimes lay down dark lines, called lines of arrested growth (LAGs), which represent periods when growth slowed or stopped for a while. LAGs are common in amphibians and reptiles, but modern birds typically lack them because they complete their growth in less than a year. In 1981, Robin Reid of Queen's University of Belfast reported that dinosaurs showed the lines, too. By counting them like tree rings, paleontologists can infer how many years a bone has grown, and by extension how long the dinosaur lived. This technique of skeletochronology is now widely used by biologists studying modern reptiles and amphibians thanks to Jacques Castanet and others in de Ricqlès's laboratory in Paris, which has trained many paleohistologists.

Interpreting fossils can be tricky. For one thing, an animal's body continually dissolves primary bone—to extract calcium or to repair microfractures—and then deposits secondary bone, erasing the bone's early history. To account for missing LAGs, researchers must make assumptions about their spacing and about bone deposition rates—no simple task, because in living animals, deposition rates vary widely from species to species and even between bones in the same individual. Temperature and diet affect bone growth, too.

One solution is to look at many specimens of various ages, so that juvenile bone fills in the missing picture for adults. "As long as I have enough individuals and a diversity of bones, I can reconstruct what was going on," says Kristi Curry Rogers of the Science Museum of Minnesota in St. Paul. By counting LAGs, researchers can assemble a series with individuals of various ages. Then, using techniques for estimating an animal's mass from the size of its bones, they plot how various types of dinosaurs typically grew over time. Most growth series are partial, but the hadrosaur *Maiasaura* is known from embryo to adult.

When growth curves were published in the 1990s, they revealed startling facts about dinosaurs. In a 1999 *Journal of Vertebrate Paleontology* paper, for example, Curry Rogers showed that the giant sauropod *Apatosaurus* reached full size—25 meters long—in just 8 to 11 years, not the decades that had long been assumed. "People would have laughed!" says Kevin Padian of the University of California, Berkeley. The quick growth rate complicates a long-standing puzzle: How did sauropods, with their relatively small mouths and simple teeth, manage to get so big, particularly during the Jurassic, when only cycads and other plants of meager nutrition were growing?

Researchers have also established a general pattern for dinosaurs, compared to other groups. In a pair of 2001 *Nature* papers, two groups—Padian's team and Erickson and colleagues—used different techniques to plot growth curves for several dinosaur species. Both concluded that dinosaurs grew faster than reptiles. The larger dinosaurs packed on weight at a pace comparable with that of mammals, but none grew as blindingly fast as modern birds do. Growth curves also show that, like birds and mammals, dinosaurs grew fast when young, then slowed down and stopped growing as adults. By contrast, nondinosaurian reptiles such as crocodiles grow more slowly.

Growth curves have been used to investigate how dinosaurs evolved various patterns of growth. In August, Erickson and several co-authors reported in *Nature* how *T. rex* evolved to its formidable size, relative to other tyrannosaurids (*Science*, 13 August, p. 930). Rather than extend its growth phase, *T. rex* accelerated its adolescent growth spurt—packing on up to 2 kilograms a day. Sauropods show similar changes, according to a paper by Martin Sander of the University of Bonn,

Germany, and colleagues, in press at *Organisms, Diversity & Evolution*. "That's not information you can get from gross anatomy," notes Allison Tumarkin-Deratzian of Vassar College in Poughkeepsie, New York. "The only way you can pin down accelerated rates of growth versus extended period of growth in the fossil record is by looking at histology."

GIANTS AND DWARFS

As a rule, dinosaurs and other vertebrates evolved to get bigger through the ages. But the bones show that some bucked the trend. At this week's SVP meeting, Sander and Octavio Mateus of the Museum of Lourinha in Portugal and colleagues announced that small sauropods discovered in Germany are not juveniles but "the first unequivocal case of dwarfing for any dinosaur." The 10 individuals range in size from 1.8 to 6.2 meters long—much smaller than the 23-meter-long brachiosaurids to which they are closely related—but the tightly spaced growth lines in their bone tissue clearly show that they were full grown. The growth curve, based on seven leg bones, suggests that the dwarfs may have been sexually mature at as young as 2 to 3 years of age.

The dwarfs lived about 150 million years ago, on an island about half the size of New Zealand. So they could provide new data about the relationship between land area and the maximum size of animals. Curry Rogers and colleagues are working on the histology of other possible "island dwarf" sauropods, titanosaurs from Argentina and Romania.

Birds are another group that reduced their body size relative to their dinosaurian ancestors, and histology is helping researchers figure out how that happened. By comparing their tissue with those of the most birdlike dinosaurs, Padian and others have argued that they shrank by shortening the amount of time they spend growing most rapidly. (Most birds reach full size within a few weeks.) "It's a very smart idea," says Luis Chiappe of the Natural History Museum of Los Angeles County. So even though they grow at a faster rate than dinosaurs, they end up smaller— which would have been a key step toward evolving the ability to fly.

Following up on research begun by de Ricqlès in the 1970s, Chinsamy-Turan is also looking at the beginnings of another fast-growing group: the mammals. Research on their therapsid ancestors shows that some of these so-called mammal-like reptiles were still growing like reptiles, while others show some distinct evidence of more mammalian growth patterns. Now she's looking at Mesozoic mammals from the Gobi desert. "I have begged and really pleaded" to get access to these rare specimens, she says, to compare them to modern mammals.

PALEO-EXOTICA

Sometimes zeroing in on ancient bones turns up exotic results. At the SVP meeting, Horner, and Mary Schweitzer and Jennifer Wittmeyer of North Carolina State University in Raleigh, described tissue, never reported from a dinosaur before, from the femur of a *T. rex*. The tissue has a random structure and is much richer in blood vessels than surrounding tissue. The researchers propose that it functioned like tissue that female birds use to store calcium for making eggshell. If so, it would be the first time paleontologists have determined gender—and reproductive status—from a dinosaur bone. Some skeptics, however, think the tissue structure might be the result of injury or disease.

Other novel tissues have been reported from flying reptiles. Pterosaur bones have plywood-like tissue made of layers stacked so that bone fibers run at right angles in alternate layers. Such crisscrossing structures are common in fish scales, but Horner, Padian, and de Ricqlès were the first to describe them in a four-limbed vertebrate. In 2000 in the *Zoological Journal of the Linnean Society*, they "speculated that the tissue is an adaptation for the biomechanical demands of flight."

Another unusual tissue has been found in squat, armored dinosaurs called ankylosaurs. Examining the bony plates called scutes, Sander and Torsten Scheyer, also of the University of Bonn, found bundles of structural fibers arranged parallel, perpendicularly, and obliquely to the scute surface—a light, strong design that would have resisted impacts from all directions, they speculate in next month's issue of the *Journal of Vertebrate Paleontology*. "It's a highly developed, composite material, like a bulletproof vest, that would prevent penetration of sharp objects," Sander says.

Histology can also be used to test hypotheses about the function of bizarre structures that no longer exist in the world. *Stegosaurus* plates have long attracted attention, and a prevalent idea is that they were used to regulate body temperature. Horner, Padian, de Ricqlès, and Russell Main, now a graduate student at Harvard University, decided to test that idea. After making thin sections of stegosaur plates and the smaller scutes of related dinosaurs, the team discovered that stegosaur plates had evolved simply by expanding the keel of scutes. "We saw nothing special about the stegosaur plates" that would be an adaptation for thermoregulation, Main says. Moreover, structures originally described as blood vessels probably weren't. The plates were probably used instead for species recognition, the group proposes in a paper in press at *Paleobiology*.

Indirectly, bone histology can even shed light on long-vanished animals' behavior. Curious about whether baby hadrosaurs would have stayed in the nest or struck out on their own after hatching, Horner's team looked at the bone tissue of *Maiasaura* embryos, as well as embryos of alligators and ratite birds. "We didn't have much evidence until we looked at the histology of the bones," Horner says. Unlike the ossified bones of alligators and ostriches, the tissue at the end of the hadrosaur limb bones consisted of calcified cartilage, suggesting that hatchlings couldn't walk immediately. They reported these findings in the *Journal of Vertebrate Paleontology* in 2000.

Walking, running, jumping, flying: Bones actively respond to the stresses of these and other physical activities. On the one hand, this can complicate the interpretation of bone tissue when researchers are trying to establish growth rates. But because physical activity affects bone, it may also be possible to extract that history from bone tissue, for example by studying the orientation of the strut-like trabecular tissue inside bones, which is often oriented perpendicularly to the major axis of strain. "It's tricky and requires a certain amount of interpretation," cautions John Hutchinson of the Royal Veterinary College in London. "There's still a lot of work that needs to be done in modern animals to see how strain impacts bone remodeling."

A good amount of that work is going on. For example, Main is studying goats to determine how biomechanics affects their bone histology. He hopes to find signals that could enable fossils to reveal posture, among other details. Other researchers are seeking similar clues in alligators, crocodiles, and birds. "Modern animals are some of the great unsung heroes of dinosaur paleontology," says Curry Rogers.

Better known heroes are playing a key role too, especially when they are abundant. Horner, for example, continues to mine a rich deposit of hadrosaurs, with individuals of all ages and sizes. "We're cutting hundreds and hundreds of slides," says Horner, who has a technician working on histology full-time. Once his group and others nail down what's normal for bone tissues, they may be able to probe the many influences that affect bone, extracting information about sexual dimorphism, climate, gait, and much else. "We've just begun to scratch the surface," says Chinsamy-Turan.

2

Bone Sharps, Fossil Hunters, and Dinosaur Experts

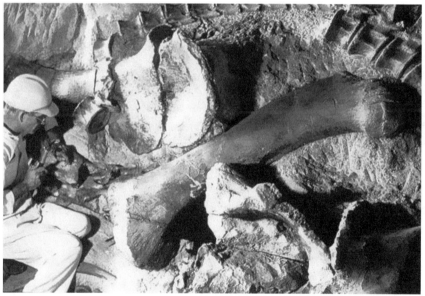

Workers examine a bone specimen now on display at the Douglass Quarry Vistor's Center, at the National Dinosaur Monument, in eastern Colorado.

This skull is prepared for study by the National Park Service and visiting paleontologists.

Editor's Introduction

Fossil hunting can be thrilling, adventurous work, but it's not every day you pull an intact *T. rex* skull out of the ground. As the articles in this chapter make clear, paleontology can also involve a great deal of waiting and drudgery.

The chapter opens with "Where Dinosaurs Roamed," which tells the story of the great American dinosaur feud between Othniel Charles Marsh and Edward Drinker Cope. Writer Genevieve Rajewski introduces the famous fossil hunters and briefly recaps their skirmishes, which played out in the field and on the pages of scientific journals. One quarry they quarreled over was home to the first fossils of the *Apatosaurus*, known for years as *Brontosaurus*. Once thought to have been "ruined" in the fight between Cope and Marsh, the site is proving to once again be a treasure trove of fossils.

Fossil beds exist all over the globe, and for almost a hundred years the Gobi Desert in Mongolia has yielded significant finds, including well-preserved dinosaur eggs. In "For Fossil Hunters, Gobi Is No Desert," the next piece, Pulitzer-winning *New York Times* science journalist John Noble Wilford describes an expedition led by paleontologists from the American Museum of Natural History. Despite powerful sandstorms and extreme temperatures, the scientists manage to make some important discoveries—among them an *Oviraptor* nest—and even enjoy homemade pizza and cold beer.

Australia is next on the world tour of fossil-hunting locales, and in "Journey to the Past," Kylie Piper of the Australian Geographic Society discusses some of the key finds made in her country. As the southernmost part of Gondwana, the lower supercontinent formed by the split of Pangaea, Australia was once home to a variety of dinosaurs. Due to its modern climate, bones can sometimes be found right on the surface, no digging required.

In "Bone Hunt: A Reporter's Week in the Wilds of Montana," the next piece, *Science News* writer Sid Perkins reports on the time he spent with paleontologists on a ranch two hours north of Billings, Montana. Perkins captures the essence of bone hunting: long periods of monotony spiked with moments of excitement. He recalls his training as an amateur assistant and his small part in excavating tail vertebrae possibly belonging to a juvenile stegosaur.

The chapter continues with "Dinosaur Shocker!" in which Helen Fields presents an unorthodox method of dinosaur study and the equally unorthodox paleontologist, Mary H. Schweitzer, credited with its creation. Schweitzer cut into dinosaur fossils in her lab and found remnants of soft tissue, effectively linking biology and paleontology and paving the way for future discoveries about dinosaurs.

This chapter ends as it began—with a conflict over the discovery of dinosaur fossils. In "The Dino Wars," *Smithsonian* journalist Donovan Webster examines the controversy spawned by independent prospectors and poachers catering to private collectors of dinosaur fossils. With the potential for million-dollar profits, the race is on to dig up more and more, regardless of the risks.

Where Dinosaurs Roamed[*]

By Genevieve Rajewski
Smithsonian, May 2008

Othniel Charles Marsh and Edward Drinker Cope were the two most prominent dinosaur specialists of the 1800s—and bitter enemies. They burned through
money, funding expeditions to Western badlands, hiring bone collectors away from
each other and bidding against one another for fossils in a battle of one-upman-
ship. They spied on each other's digs, had their minions smash fossils so the other
couldn't collect them, and attacked each other in academic journals and across the
pages of the New York Herald—making accusations of theft and plagiarism that
tarnished them both. Yet between them they named more than 1,500 new species
of fossil animals. They made *Brontosaurus*, *Stegosaurus* and *Triceratops* household
names and sparked a dinomania that thrives today.

One of Marsh and Cope's skirmishes involved fossil beds in Morrison, Colorado, discovered in 1877 by Arthur Lakes, a teacher and geologist-for-hire. Lakes
wrote in his journal that he had discovered bones "so monstrous . . . so utterly
beyond anything I had ever read or conceived possible." He wrote to Marsh, at
Yale, to offer his finds and services, but his letters met with vague replies and
then silence. Lakes then sent some sample bones to Cope, the editor of *American
Naturalist*. When Marsh got word that his rival was interested, he promptly hired
Lakes. Under Marsh's control, the Morrison quarries yielded the world's first fossils
of *Stegosaurus* and *Apatosaurus*, the long-necked plant eater more popularly known
as *Brontosaurus*.

Lakes spent four field seasons chiseling the most easily reached bones out of the
fossil beds. Before he left the area, he allegedly blew up one of the most productive
sites—"Quarry 10"—to prevent Cope from digging there.

For 123 years, the site was lost, but in 2002 researchers from the Morrison
Natural History Museum used Lakes' field notes, paintings and sketches to find
the quarry, expose its original floor and support beams and begin digging once

* Article by Genevieve Rajewski first published in *Smithsonian Magazine*, May 2008. Reprinted by permission of the author. All
rights reserved.

more. "The first things that we found were charcoal fragments: we were digging right below the campfire that Arthur Lakes had built," says Matthew Mossbrucker, director of the museum.

They quickly discovered that at least one misdeed attributed to the feud between Marsh and Cope was probably exaggerated. "It looks like [Lakes] just shoveled some dirt in there," says Mossbrucker. "I think he told people that he had dynamited it closed because he didn't want the competition up at the quarry—playing mind games with Cope's gang."

The reopened quarry is awash in overlooked fossils as well as relics that earlier paleontologists failed to recognize: dinosaur footprints that provide startling new clues about how the creatures lived.

The dig site is perched halfway up the west side of a narrow ridge called the Dakota hogback. The only way up is to walk—over loose rock, past prickly brush and rattlesnakes—with frequent pauses to catch one's breath. On this July morning, Mossbrucker leads six volunteers as they open the quarry for its fourth full modern-day field season. The crew erects a canopy over the pit before forming a bucket brigade to remove backfill that has washed into the hole since last season.

Down in a test pit, the crew digs into the side of the ridge, carefully undercutting the layer of cracked sandstone that served as the original quarry's ceiling. The ledge collapsed several times in the 1870s. More than 100 tons of rock crashed into the pit one night, and had the crew been working instead of sleeping nearby, Lakes wrote, the "entire party would have been crushed to atoms and buried beneath tons of rocks which afterwards took us over a week to remove by blasting and sledge hammers."

Robert Bakker, curator of paleontology for the Houston Museum of Natural Science, helps out at the dig. "If you want to understand the late Jurassic, you need to understand the common animals, which means *Apatosaurus*," he says. "This is the original *Apatosaurus* quarry, and it's a 'triple-decker'—the only one in the world with three dead *Apatosaurus* buried one on top of each other."

Most people know *Apatosaurus* as *Brontosaurus* because of a mistake made by Marsh. In 1879, two years after he named the first *Apatosaurus*, one of his workers discovered a more complete specimen in Wyoming. Marsh mistook it to be a new animal and named it *Brontosaurus*. Though the error was soon discovered, scientific nomenclature required keeping the first name. But in the meantime the "*Brontosaurus*" misnomer had made its way into popular culture.

For almost 100 years, *Apatosaurus* was portrayed as a swamp-bound animal whose immense body was buoyed by water. In the 1960s, Bakker joined a handful of paleontologists who argued that the massive beasts were really more like elephants: all-terrain animals that could roam over the flood plain, through river channels and anywhere else they wanted to go.

Bakker, then an undergraduate at Yale, went to Morrison to see if *Apatosaurus'* habitat supported his idea that the beasts were mobile. But he and two students spent two years unsuccessfully hunting for Quarry 10, which aside from being partially filled in, as Bakker finally discovered, was also obscured by bullet cartridges, beer cans and other remnants of teenage outings.

Today, Bakker is sifting through Lakes' spoil pile —lumps of clay stone that the 1870s crew tossed aside—when someone in the pit excitedly calls for him. He scrambles down into the hole, where his bearded face lights up under his straw cowboy hat. The museum crew has uncovered what appear to be Jurassic-era castings of a small tree's root system. "This is a big deal," says Bakker, using a finely bristled brush to baste the knobby fossils with glue. "In 'CSI' terms, that's the crime scene floor. Victim number one"—the *Apatosaurus* found in 1877—"lay buried just above."

The clue adds to evidence that *Apatosaurus* did not live in water. The team has found layers of sediment consistent with a small pond, but none of the crocodile or tortoise fossils typically found in swamps from the Jurassic Period more than 200 million years ago. This spot may have attracted generations of *Apatosaurus*, Bakker says, because it provided a watering hole on a dry wooded plain. "If there was a forest, there would be a lot more wood—and there isn't—and a lot more fossilized leaves—and there ain't. So it was a woodland but probably a lot like Uganda—hot tropical woodland that was dry for most of the year."

The most significant recent discoveries at the Morrison quarries have been dinosaur tracks. Early dinosaur hunters overlooked them. In Quarry 10 and another Lakes quarry less than a mile away, museum staff have recovered 16 *Stegosaurus* tracks. They include ten hatchling tracks—the first ever discovered. One rock appears to show four or five baby *Stegosauri* all heading in the same direction. Another boulder includes a partial juvenile *Stegosaurus* hind paw track that was stepped on by an adult *Stegosaurus*. "It suggests that *Stegosaurus* moved in multiple-age herds," says Mossbrucker, and adults may have cared for hatchlings.

The researchers have also found the world's first baby *Apatosaurus* tracks. They could change paleontologists' view yet again: the tracks are from the rear legs only, and they are spaced far apart. "What's really cool about these tracks is that the baby animal is functionally running—but it's doing this just on its back legs. We had no idea a Bronto could run, let alone scoot along on its hind legs like a basilisk," Mossbrucker says, referring to the "Jesus lizard" that appears to walk on water.

He and others speculate that adult Apatosauri, some of the largest animals ever to walk the earth, could prop themselves up on two legs with the help of their long tails. But others argue that it would have been physiologically impossible to pump blood up the animals' long necks or to raise their heavy front limbs off the ground.

Bakker and Mossbrucker say their goal is to look at Quarry 10 holistically—considering the local geography, climate, flora and fauna—to create a picture of where and how Jurassic dinosaurs lived. "I want to know as completely as I can what kind of forgotten world these dinosaurs knew," says Mossbrucker. "I want to see what they saw, touch their earth with my own feet and be in the Jurassic."

Bakker gestures toward the pit, where Libby Prueher, the museum's curator of geology, sifts soil alongside volunteer Logan Thomas, a high-school student with a passion for snakes. "It's weird that [Marsh and Cope] thought that dinosaurs were a zero-sum game, that Marsh thought, 'If Cope got a bone, I lost a bone,'" says Bakker. The goal isn't to vanquish one's rivals, he says: "the guiding inspiration for studying the dead dinosaurs is to get back to how they lived."

GENEVIEVE RAJEWSKI, *a Boston-based writer, caught dinomania as a child and is surprised by how much paleontology has changed.*

For Fossil Hunters, Gobi Is No Desert[*]

By John Noble Wilford
The New York Times, September 13, 2005

On the first afternoon here [Ukhaa Tolgod, Mongolia], fossil hunters struck out across the parched sand to the rock outcrops along the bleached brown ridges and down into the broad basin. They walked their separate courses at paces as if set to geologic time.

With every step, their figures diminished into the expanse of empty silences and far horizons that is the Gobi Desert, where only camels, nomads and hardy paleontologists seem at home.

It has been the paleontologists' boast, never disputed, that this particular forbidding stretch of the Gobi holds the world's richest and most diverse deposits of dinosaur and early mammal remains from 80 million years ago, a critical time for life in the Cretaceous geologic period.

Four years had passed since paleontologists of the American-Mongolian expedition last pitched camp at Ukhaa Tolgod ("brown hills" in Mongolian), scene of their greatest triumphs. They were lured back last month, as surely as gold prospectors to the mother lode, by the expectation that the site has more to yield.

The figures in slow motion appeared indifferent to the baking heat. Each one assumed the prospecting posture: head down, swiveling from side to side, eyes fixed on the ground. At the sight of a fleck of white or an arresting irregularity in the ground, a figure bent for a closer look, then tossed something over the shoulder and wandered on, a common experience in fossil hunting.

The badlands of Ukhaa Tolgod were still generous. Clues in the ground frequently brought the prospectors to their knees. They scraped and dug the hard-packed sand with knife, chisel or geologist's hammer.

If one of them stretched out as if to take a nap, here of all places, it was to apply more delicate strokes and a more discerning eye to the task. This is the discovery posture, and the longer the prone position is maintained, the more likely the hunter will be going back to camp with an exciting find.

Nearly everyone this afternoon had something new to talk about. The haul included two nests of dinosaur eggs, the remains of several ancient lizards and the skull and skeleton of a small mammal, most likely a previously unknown species that lived in the shadow of the formidable dinosaurs.

"Sometimes you go days, whole expeditions, without finding something," said Gina D. Wesley-Hunt, a paleontologist at the Smithsonian Institution. "We found 10 skulls just today, early mammals and lizards."

Happy and relieved by the finds, Michael J. Novacek, a paleontologist at the American Museum of Natural History who has directed the expeditions since their start in 1990, declared, "This place is still rich."

One heard in his words an echo from the celebrated Gobi explorations of Roy Chapman Andrews, also of the American Museum in New York City. At one of the team's first stops, in 1922, a geologist rushed into camp to announce, "The stuff is here."

Andrews put the Gobi on the map as a remote land of fossil wealth with his discoveries and his vivid prose. The desert extends more than 1,500 miles across southern Mongolia and reaches into northwestern China. Russian, Polish and Mongolian scientists made fossil-hunting forays here after World War II. But not until Mongolia declared its independence from the Soviet Union were Americans invited back. In 1993, after three somewhat disappointing seasons, they made their "killer find" here.

In the first three hours, the party extracted from the nearest slopes fossils of 60 dinosaurs, mammals and lizards, an unimaginable rate of one find every three minutes. It was an easy decision to stay another 10 days. In that time, the paleontologists found more dinosaurs—oviraptors and troodons—and a virtually complete skull of a Mononychus, which is thought to be a transitional species between dinosaurs and modern birds.

The most poignant discovery was a nest of dinosaur eggs with a broken one exposing a fossilized curled-up embryo. Something never before seen, it is now a museum piece referred to as the embryo on a half shell.

Return visits to Ukhaa Tolgod through 2001 brought other striking finds. Then the scientists decided to give the place a rest and see if a few years of stiff wind and sporadic rain eroded away a fraction of an inch of sand and rock, possibly enough to bring to light more goodies. They were right; there was stuff still here.

DESERT LUXURIES

In camp that evening, the beer was cold. Under other circumstances, this observation would go without remark. Beer is supposed to be cold, just as the Gobi wind is supposed to kick up the sand, starting in late afternoon, and redistribute it into your tent, your eyes and the food you eat.

But this cold beer had refreshing meaning for Dr. Novacek, who is also vice president and provost at the museum, and Mark A. Norell, the museum's principal dinosaur paleontologist and field leader for all but one of the expedition's 16 seasons. It was a surprise and delight for this reporter.

I had accompanied Dr. Novacek and Dr. Norell for three weeks on their first full season in the Gobi, in the summer of 1991. It was a no-frills reconnaissance for the most part over spine-jarring terrain Andrews and others had never explored.

We did some prospecting in the red sandstone at many places and kept moving camp in constant search of promising dig sites. We had no communications with the outside world. Our rations ran short, and it had been a miscalculation to count on replenishment in the rare settlements we came to. We subsisted mainly on mutton, purchased every few days from any nomad herder we happened on.

And the beer after a hard day fossil hunting was warm, and too soon beer at any temperature was a memory.

I joined this year's expedition with a Times photographer, Chang W. Lee, two officials of the American Museum and a crew shooting a television documentary. I wanted to live again in the harsh grandeur of the Gobi, inspect this treasure-chest fossil site I had heard and written about from afar and see if the grinding field work of paleontology had changed much in the intervening years.

We arrived from Ulan Bator, the capital, by helicopter, a flight of some 500 miles and much more comfortable than the alternative of a three-day journey by truck over rough roads, sometimes no roads at all. The flight took us south of the rolling grasslands into increasingly barren territory. The Gobi itself had not changed.

The land at first is corrugated with low ridges and narrow valleys. Soon, the Altai Mountains in the distance, two ranges running widely parallel east to west, the southern one bordering China, are an ever-present backdrop in subtle tints of vermilion, mauve and khaki.

Farther west, the desert plain between the mountains opens wider and browner. Its flatness is relieved here and there by dry streambeds and gullies, low hills and brown ridges capped with a veneer of gray gravel.

Sand and more gravel are everywhere, and a scattering of dry bush and only infrequent patches of stubby grass being grazed by horses, two-hump Bactrian camels, sheep and goats. It is miles and miles between encampments of gers, the nomads' traditional round, canvas-covered, felt-insulated mobile homes.

Ukhaa Tolgod is about 250 miles southwest of Dalandzadgad, a provincial capital and one of the few towns of any size in the Gobi. The 13-person science party, which had already been in the field working other sites for more than two weeks, had arrived the night before their visitors. The team's yellow tents, a ger and parked trucks stretched out at the foot of a high ridge known as the Camel's Humps.

When the subject of cold beer came up, a grin creased Dr. Norell's bearded face. "Yeah, different, isn't it?"

Dr. Novacek pointed to a large array of solar-energy collectors. The camp had electricity to charge batteries, computers and limited communications and to keep food and beer refrigerated. The expedition had more trucks than before for hauling in gasoline, water, field equipment and an ample supply of food as well as cans of beer.

It was not gracious living, but there were fewer grounds for complaint than on earlier ventures into the Gobi.

AN ABUNDANCE OF MAMMALS

No one was complaining about anything at the end of the first afternoon's prospecting. The camp buzzed with show-and-tell chatter. Attention centered on Julia Clarke's find, the well-preserved skull and some skeletal bones of a mammal that might have been the size of a small rabbit.

"For a brief moment, I thought it might be a bird," said Dr. Clarke, who teaches paleontology and evolution at North Carolina State University and is a researcher at the North Carolina Museum of Natural History, both in Raleigh. Her specialty is early birds.

"Then I saw the snout and huge teeth," she continued. "Early birds had teeth, but they were not huge. I will keep trying for a bird."

By now, the delicate fossils were laid out on a cloth over the ground. Two men sprawled beside the specimen, repeating the discovery posture. Their faces were inches from the skull that intrigued them.

Guillermo W. Rougier, an Argentine paleontologist on the anatomy faculty at the University of Louisville School of Medicine, held a small magnifying lens to the head and picked at it with an awl. "One of the best skulls we have," he said.

He and Dr. Novacek, lying on the other side, thought it could be a little marsupial, perhaps an ancestor to lineages leading to kangaroos and opossums. If so, it would support the hypothesized Asian origins of the marsupials that eventually migrated to Australia and the Americas.

No, the teeth looked more like those of an early eutherian, ancestors to modern placental mammals, the group that includes humans. Placental mammals bear live young after a prolonged pregnancy. This particular early eutherian occupied the ecological niche of rabbits and had the long hind legs of rabbits and the habits of rodents, but was unrelated to either.

Dr. Rougier began applying gauze and strips of burlap soaked in wet plaster around and over the specimen, preparing it for the trip to a laboratory in something like a broken-leg cast.

"It isn't like anything we've found here before," Dr. Novacek said. "We really won't know what we have until we get it back to the lab."

For every month in the field, paleontologists know from experience, the study of fossils requires at least another 11 months of tedious indoor work back at their universities and museums. But one of the immediately evident changes in Gobi research is the increasing number and variety of small mammals turning up in excavations.

This development is expected to bring new insights into the formative epoch of mammalian life in the Cretaceous Period, which lasted from 140 million years ago until the extinction of nonavian dinosaurs 65 million years ago.

"When you read some accounts, you would think only dinosaurs lived here in the Gobi," Dr. Novacek said, a while later. "That drives us crazy. The Cretaceous was a fantastic time in Earth history. The world we live in began in the Cretaceous, the modern ecosystem, flowering plants and pollinating insects, the origin of modern mammals."

In the fossil beds at Ukhaa Tolgod, laid down in the late Cretaceous and entombing a broad sampling of its life, not only dinosaurs, paleontologists estimate that over recent years they have found 1,000 mammal skulls. Dr. Novacek said this amounts to 90 percent of all the recovered mammal specimens from the Cretaceous. The beds have also yielded remains of 1,000 lizards. Several of their distant descendants, in sand-colored camouflage, skittered before my steps to the tent in the Gobi night.

THE DOCTOR'S DINOSAURS

Next morning, Dr. Norell drove an S.U.V. to the hillside where he had come upon a dinosaur nest the day before. From the slope, he looked over to the saddle between two hills, where in 1993 an expedition truck got stuck in deep sand. The scientists took the mishap as an opportunity and made the first of their findings revealing the site's richness.

Three eggshell fragments on the surface had led Dr. Norell to his new find. The smoothness of the shell, he said, indicated that it was from an egg of a member of the troodontid group of dinosaurs, possibly a Byronosaurus. Digging carefully in the sand, he exposed a nest of broken eggs. The tiny animals had presumably broken open the four-inch-long oval eggs and climbed out. None of their fossils were present, but the paleontologist looked forward to making a detailed study of the shells.

"If you found this anywhere else in the world, you'd be going crazy now," Dr. Norell said, while freeing the nest from the sandstone. "But it takes a lot to top our best stuff, like dinosaurs sitting on top of nests and eggs with embryos, and some other things we have not published on yet."

Later, paleontologists went to examine another nest, one that Dr. Matthew R. Lewin spotted the first afternoon. Dr. Lewin was the expedition physician, a position unheard of in the earliest seasons.

A medical problem brought him to fossil hunting in the last three seasons. For years, some expedition members had complained of persistent fevers, rashes and swollen lymph nodes. Dr. Lewin, an emergency-room doctor and researcher in physiology at the University of California, San Francisco, studied the cases and determined the cause was a tick-transmitted disease, like Rocky Mountain spotted fever, which was readily treatable with antibiotics.

One thing led to another and he signed on as the expedition doctor.

"Finding a nest is really fantastic," Dr. Lewin said, as he and Dr. Norell prepared it for removal. "I saw one shell fragment on the ground. I played the game of hot-and-cold. I walked in one direction. It was cold, no shells. I walked in another direction and saw several shells. I was getting hotter. Digging around, I saw five eggs on top and at least two on the bottom, maybe more."

Dr. Norell identified the physician's fossils as an oviraptor nest, judging by the oval shape and the size of each egg, six to eight inches long, and the shell texture, with its minute ridges.

Oviraptors were carnivores, the top predator in the food chain here. From these eggs they grew to lengths of eight feet, had curved claws and crested skulls and walked, ostrichlike, on their hind legs.

It is a good bet that, on his return to San Francisco, Dr. Lewin will talk more about his oviraptor nest than the few wounds, infections and dehydration cases he treated while in the Gobi.

SUN AND WINDSTORMS

Nights in the Gobi are resplendent under the bright arc of the Milky Way and are usually cool, especially here in the high desert, about a mile above sea level. Morning light in summer breaks after 6 a.m. In no time, tents become sweltering saunas and there is nothing to do but emerge into the daylight under an ascending sun.

Add fossil hunters to the list of mad dogs and Englishmen who go out in the noonday sun. Temperatures reach 100 degrees Fahrenheit nearly every summer day and become almost unendurable by late afternoon. There is no natural shade, except in the shadow of a rock face. Yet the scientists went out for several hours each morning, and back again for most of the afternoon, prospecting or excavating.

Without fail, a crew sweated out the heat high on the sandstone cliff of the Camel's Humps. They worked with the full complement of excavation tools: picks and shovels, sledges and chisels, whisk brooms and the ever-handy geologist's rock hammer. Dr. Novacek calls the hand-held tool, with a small pick opposite the hammer head, "the Excalibur of the paleontologist."

The crew usually included Dr. Rougier, Dr. Wesley-Hunt, Pablo Puerta, an Argentine museum technician, and Andres Giallombardo, an Argentine graduate student at the American Museum and Columbia. Bending to their work on a ledge, they extracted two of the five juvenile dinosaur skeletons, which had been left behind, reburied, at the end of a previous visit.

These were ankylosaurs, probably a Pinacosaurus, abundant plant eaters and the prey of predatory dinosaurs. If the juveniles, about 6 to 7 feet long, had lived, they would have grown into 25-foot-long spike-tailed adults.

In late afternoon, dark clouds often move in and the wind picks up. It usually blows fiercely for a couple of hours, testing the stakes and lines of tents and scattering anything not tied or weighted down. A raging sandstorm on my first Gobi trip uprooted my unoccupied tent and sent it bounding several hundred feet across the desert, until we could finally catch up with it.

The scientists with long Gobi experience insisted that the winds now were relatively tame. But one storm persisted longer than usual. The visitors hunkered down in their dining ger, a short distance from the science camp. A separate camp had been set up for the visitors, with amenities supplied by a commercial expedition outfitter.

As the wind shook the ger, those from the plains of middle America got to talking about tornadoes they had weathered. An Oklahoman told an old joke: "You know what a divorce and a tornado have in common? In either case, someone loses a mobile home."

Scientists once thought that violent sandstorms accounted for the number and excellent preservation of fossils hereabouts. The storms buried the animals alive in their tracks and burrows, an act so sudden and devastating that the sand-covered bodies were never scavenged and had undisturbed time for their bones to mineralize as fossils. Now, a more complex natural phenomenon is considered likely.

When a sandstorm is accompanied by torrential rain, which sometimes happens, the water-saturated sand dunes in an earlier time could have collapsed and buried life all around. Scientists suspect this is what killed the juvenile ankylosaurs on the cliff, where the sandstone is the lithified remnant of an ancient dune.

So Ukhaa Tolgod was not only a Cretaceous nursery of nested life, but also a valley of death. Storms fiercer than the one blowing now are what keep paleontologists in business.

MONGOLIAN HOSPITALITY

Dr. Novacek said that next year he wanted the expedition to concentrate on the desert's geology. An effort will be made to date more precisely the age of known sites and establish their temporal relationship to one another. Also, research will be conducted on the minerals in the rocks, determining their oxygen content.

Minerals hold markers of oxygen levels in the atmosphere at the time of their formation. The air today is about 21 percent oxygen. In the Jurassic, the geologic period before the Cretaceous, it was as low as 12 percent, which Dr. Novacek said would have been hard on large animals.

About 100 million years ago, in the middle Cretaceous, the atmospheric oxygen levels started rising, until the present condition was reached 50 million years ago. This could be a factor in the flourishing of mammals after the dinosaur extinction 65 million years ago.

"This is an exciting area of paleontology today," Dr. Novacek said. "We don't know yet how oxygen levels exactly relate to organisms, but this is bringing physical data to bear on life science in studies of biota and ecosystems."

But hunting for other sites for future fossil exploration was on the mind of some of the expedition scientists, who took advantage of the helicopter to make daily scouting runs across the desert.

On one trip, deep in the Nemegt Valley, Julia Clarke descended into a deep canyon. It was good to get out of the hot sun. She scanned the sandstone walls and the ground underfoot. Walking slowly, expectantly, she said she was looking for "any tiny flecks of white, bone fragments that are often bluish white."

Mark Norell scouted a maze of gullies. He spotted the scat of the wild Marco Polo sheep and a set of one's horns. Then the prints of a wolf, a hedgehog and a jerboa, a kangaroo rat. But no fossils. All he could say was, "The Polish worked here years ago and found quite a lot."

Demberelyn Dashzeveg, a 74-year-old paleontologist at the Mongolian Academy of Sciences who has a keen eye for fossils, had more luck in another canyon, known as Zos and only six miles from camp. The geology of the site indicated that this had been an oasis 80 million years ago, the shales and mudstones suggesting there had been standing water. He found bones and teeth of a possible new mammal species.

On the return from a more distant site, Namsrai, the driver (who like most Mongolians uses only his given name), steered the Land Rover with skill and audacity over the rough desert flats, bouncing us up and down and side to side to the Motown music blaring from the tape deck.

We came to a nomad's ger. A black dog guarded the wooden door, but it was friendly enough, as were the two young girls who opened the door and invited us in.

Mongolian nomads are noted for their hospitality. The girls, probably 7 and 8 years old, motioned for us to sit on the stools and cots around the metal stove in the center, its flue rising through an opening at the apex of the domed ger. Hanging on the tent wall, at the back, were equestrian medals the father had won at Nadam, the country's annual festival of horse racing, archery and wrestling.

The girls busied themselves setting out cups and getting a thermos from a cupboard. They poured warm tea with sheep's milk and served a plate of breadsticks. Alice in Wonderland could not have hosted a finer tea party.

The father was out on his horse tending the herds. After a while, the mother appeared. She had been at the only neighboring ger, probably watching television. Solar panels and a satellite dish, also a dirt bike and a pickup, could be seen at the other ger. Some changes in life had come to the Gobi.

PLEASURES OF FIELDWORK

Evenings in camp often end around an open fire, its warmth, the gathering dark and their weariness bringing the scientists together. The treat one evening was pizza. Dr. Clarke prepared the toppings of cheese, tomato sauce, mushrooms and salami—and the inescapable sprinkling of sand. Dr. Lewin baked the pies in two Dutch ovens over smoldering embers.

Another time, the scientists sat in the sand around a blazing fire, passing a bottle of Genghis Khan vodka for communal swigs. They talked of past discoveries, academic politics back home, memories of hot-water baths and the howling of a wolf they heard the other night. Wolves, heard but not usually seen, are sources of Gobi versions of ghost stories as the campfire burns low.

"Paleontology is not a science rich in money," Dr. Novacek was saying. "Our comforts in the field are limited. But the reality is, we like it that way. We prefer to live a little on the edge."

Fossil hunting in the Gobi of Mongolia has not really changed in the last 15 years, not in its essentials, except for the cold beer.

Journey to the Past*

By Kylie Piper
Australian Geographic, July-September 2010

I was flat on my stomach in the dirt, feet in the air, when I heard dig leader David Elliott yell: "Don't you bloody well find anything else down there." But I had.

It was August 2007, we'd been digging for two weeks and this was the last bone to come out of the ground. It was a gastralia from a carnivorous dinosaur, a small floating rib that looked more like a chopstick than a piece of prehistoric giant. We'd run out of plaster and hessian—the first line of defence against the atmosphere from which bones have been safe for millennia—and had to get off the property before shearing began. So, I climbed out of the hole and carefully wrapped the fragile piece in paper and foil.

On my first dig in Winton, Queensland, five years earlier, we'd been excited by any tiny, weathered bone fragment we found, and diligently collected and mapped each one in the hope that it might lead to a larger trove below. I volunteered to dig again each year after that, but without much success. But things started to change in 2006, when the owner of the property on which we were digging came to visit and mentioned another site she knew of. I was with Queensland Museum palaeontologists Dr Scott Hocknull and Dr Alex Cook when she showed them the site that yielded 'Matilda'—Australia's most complete species of sauropod from the Cretaceous. "I don't think we really ever thought that something so well preserved could be hidden underneath the black soil," Scott says.

Nothing moves quickly in central Queensland, especially in the world of palaeontology. It took three more years of digs and two years of preparation work before those bones were presented in June 2009 as *Diamantinasaurus matildae*, a herbivorous giant of a sauropod 16 m long and 3 m tall at the hip. But perhaps most astonishing was what lay beneath Matilda. After her huge bones were removed, the smaller, elegant remains of a hitherto unknown theropod were found tucked safely below. 'Banjo' (*Australovenator wintonensis*), with his 30 cm long claws and flesh-slicing teeth, stole the spotlight and is Australia's most complete carnivorous dinosaur. "We were used to big sauropod bones by now, but the discovery of another very different dinosaur from the site blew me away," Scott says. "Australia's dinosaur world was right there under our noses all along."

Australia's fossil beds are in some of the most inhospitable and remote regions of the nation; from the daunting Otway Ranges of Victoria's southern tip, to the underground opal mines of Lightning Ridge in north-west NSW and the blacksoil plains of central Queensland. And the usual rules of palaeontology don't seem to apply in Australia. The best fossils overseas are found in geological layers that have been folded or eroded over millennia to display vertical beds from which the fossils protrude like some new-age artwork. Australia's ancient landscape has undergone little movement; the layers of time have remained mostly horizontal, exposed only by harsh weathering. It's thought that, in the rest of the world, only about one third of all dinosaur species that existed have been discovered. Down Under, we're likely to have found far fewer than that.

David Pickering is collection manager of Museum Victoria's (MV) fossil vertebrates and plants. He pulls out drawer after drawer in the climate-controlled warehouse to show me hundreds of boxes, each with a bone from ornithopods and ankylosaurs, and tiny little vials with theropod teeth. Their serrated edges are still sharp, after 100 million years in the earth.

The museum holds an astonishingly varied range of species, including the remains of small herbivorous Victorian ornithopods with tongue-twister names such as *Fulgurotherium* and *Atlascopcosaurus*. Many were discovered by palaeontologist Dr Tom Rich, a team of volunteers and his wife and colleague Professor Pat Vickers-Rich of Monash University, during the past three decades, and prepared lovingly by David and Lesley Kool at MV.

Size isn't everything. One of the most prized possessions in the Victorian collection is the skull of *Leaellynasaura*, a tiny herbivore with giant eyes and optic lobes—thought to be a physical adaptation to the dim light in their cold, forested polar environment. During the time of the dinosaurs, Australia was part of Gondwana, and its coastline stretched from subtropical zones in the north to polar latitudes in what is now Victoria. Another species, *Timimus*, is thought to be a small ornithomimid (or "bird mimic"), and these ostrich-sized, long-legged dinosaurs were some of the fastest runners of all. The species was described from a single thigh bone.

The work of Tom, Pat and their team has more than doubled the tally of Australian dinosaur species. Together they have named many of Australia's dinosaurs, including *Timimus* and *Leaellynasaura*—named after Tom and Pat's son and daughter. But the identity of many bones remains a mystery. The 2010 announcement in the journal *Science* of an as-yet-unnamed tyrannosauroid—a tyrannosaur-like dinosaur—has caused the most recent excitement. It was described from a single hip bone first discovered in 1989 and safely stored in MV's drawers for two decades. One-third the size of *T. rex* and 40 million years older, this carnivore is the first tyrannosauroid known from the Southern Hemisphere, and shows these animals were much more widespread than first thought. Australia's tyrannosauroid possibly had feathers, as did other members of the group.

The bones of *Timimus* show some evidence of hibernation, another coping mechanism for polar conditions—and another clue about where to start looking for more dinosaurs. In fact, a 2005 discovery in the USA suggested we might have been looking in the wrong places. Tony Martin, an animal-tracking expert at Emory University, in Georgia, discovered small ornithopods in Montana in what appear to be prehistoric burrows. An adult and two juveniles of the species *Oryctodromeus cubicularis* ("running digger of the lair") were found together in the 95-million-year-old den. Tony's research opens [. . .] new prehistoric possibilities, and on a trip to Australia last year he identified two possible burrows on Victoria's coast.

It's a beautiful morning looking out towards Bass Strait, although the sun is yet to climb above the horizon. I've taken the winding coast road to the Bunurong Marine Park, near Inverloch on the southern Victoria coast, which is home to a 17-year-old fossil site dubbed Dinosaur Dreaming. Each summer, a motley and enthusiastic team of volunteers convene here to search for the dinosaurs that trod the icy forests of southern Gondwana more than 110 million years ago (mya). And every day they race against time. As the tide falls, the palaeontological chain gang begin the painstaking task of cracking open rocks. Each piece is cut down to the size of a sugar cube and carefully probed for hints of bone.

Before midday the water starts to slowly creep back towards the shore, encroaching on the dig site. I notice a trickle snaking down into the pit, just as someone discovers something. The pace suddenly picks up—now it's a battle against the tide. The volunteers are ankle-deep in water, but the bone—possibly a piece of armour from a plant-eating ankylosaur—is stuck too deep to be easily removed. Even sandbags can't stop the tide and after all that careful work, the decision is made to rebury and protect the bone before sand and seawater swallow it once more. The sea is both friend and foe, making the site difficult to access but, at the same time, protecting it from fossil thieves.

"It's persistence," says Tom Rich of dinosaur hunting. "Both luck and persistence; that's what pay off in this business." Tom first took up the quest in 1978. His mission was to find ancient mammals; species that might help to fill the gaps in Australia's largely unique marsupial record. "The thing about Australian dinosaurs is that we haven't found the dinosaur equivalent of a koala," Tom tells me, his American accent still strong. "There're no dinosaurs recognised so far in Australia that are really distinct to the known dinosaur families in the world."

The most significant discovery recently is *Sinosauropteryx prima*, the first known feathered species. Found in Liaoning Province, China, in 1996, it was the definitive evidence that birds descended from dinosaurs. Similarities between dinosaurs and birds have long been noted. Both have a characteristic wishbone and hollow bones. *Archaeopteryx*, a primitive bird with a bony tail and teeth, was key evidence in early evolutionary debates after its discovery in 1861. It's now commonly believed that many dinosaurs had downy coverings or even fully developed flight feathers. Tom sees similarities between the deposits of Koonwarra, Victoria, and those in Liaoning, which are of similar age. About six feather fossils have been found at Koonwarra. "The rocks in the [Liaoning] site look so much like the stuff at Koonwarra," Tom says. "It's a fairly unusual thing to do—to preserve a feather—and Koonwarra does that."

The problem is getting to the proof. "The most interesting dinosaurs are not all that common in [Liaoning]—but the reason they get them is because they clear an awful lot of real estate," says Tom, who estimates that more than 50 sq. m of solid rock needs to be removed in Koonwarra before anything of note is found. Only 20 per cent has been shifted so far.

Tom's plan for the next few years is to scour the coastline of Victoria and map any occurrences of what may be burrows. "We're trying to find a place where there're burrows exposed, so we can get at them. The idea would be to dig a few of them out and see if you've got any bones in them," Tom says. Unfortunately, the two possible burrows that Tony Martin identified are in a place that would make excavating a tunnel to China look easy.

In 1978 university students and cousins John Long and Tim Flannery travelled to Eagles Nest, Victoria, to hunt dinosaurs. Australia's first dinosaur fossil—the Cape Paterson Claw—had been found there in 1903 by William Ferguson, a geologist with the Mines Department of Victoria, who was following the coal seams that thread the region.

"I'd been collecting fossils since I was seven, but I'd always wanted to find a dinosaur," says John, who has studied Australian palaeontology for more than three decades and is based at the Natural History Museum of Los Angeles County, in California, USA. "We scrambled down the cliff; it was a grey rainy sort of a day, and within about five minutes of getting to the beach I'd picked up a rock that had a bone running through it . . . a dinosaur bone." Soon after, they returned to the area and found another rock, which held the femur of a small herbivorous ornithopod, and was the catalyst for the sustained hunt for dinosaurs in Victoria that continues today.

The majority of Australian species of dinosaur are described from just a single bone. In some cases, even well-researched bones are giving up new information. "With increasing technology we can go back to old fossils and get more of their secrets out of them," says John. "And that's been the success of our last five years of research with Devonian fish. We've even been able to go back to specimens we described 20 years ago and find things out about them that we never dreamt of. It's an exciting time for Australian palaeontology. [Matilda and Banjo] were the first big discoveries in about 30 years in Australia to really rock the boat."

More momentous finds are not only possible but likely, according to John. "For those looking, there are whole mountains in Western Australia that are Cretaceous and are of the right rock types to find dinosaurs in them," he says. "The other rocks in that Kimberley region of the right age have also produced fossil vertebrates of the Triassic, so we know that it's just the right setting to find dinosaurs, but it's just so hard to get out there . . . I reckon that's where the really big discoveries will be made one day."

On my first dig I found nothing except a handful of blisters, but I was hooked. Across Australia those bones are there, waiting for someone with the right combination of curiosity, determination and ingenuity to mark "X" and start digging, continuing the search for Australia's dinosaur dreaming and the chance to write more prehistory.

Bone Hunt*

A Reporter's Week in the Wilds of Montana

By Sid Perkins
Science News, August 26, 2006

"I think I've found something!" The call rang out from across the quarry. Suddenly, a dozen or so would-be paleontologists—myself included—shifted their mental focus from the small zones of rock immediately in front of them to a new center of attention. Having spent the last few hours using hand tools to grub our way through crumbly rock with little tangible result, we found the idea that someone had actually found something to be exciting indeed.

Nate Murphy, the paleontologist in charge of the dig, strolled over to take a look. "That's something, all right," he said. A little more excavation revealed the 3-centimeter-long tip of a theropod dinosaur's tooth. Considering the age of the rocks that entombed it, Murphy estimated that the meat eater had shed the fragment around 150 million years ago.

This episode, the first thrill on my recent foray into paleontology fieldwork, was by no means the last. Sure, most of those thrills were vicarious. Other folks found many more fossils—and more impressive ones—than I did. Nevertheless, I gained an understanding invaluable to my writing about paleontology—how dinosaur bones start their journey from rock formations into museums.

My invitation from Murphy, research director of the Judith River Dinosaur Institute in Malta, Mont., came late last year. "Have you ever been on a dinosaur dig?" he suddenly asked during a chat at October's annual meeting of the Society of Vertebrate Paleontology. "You need to understand what goes on in the field."

As I'd already suspected, extracting fossils from their stony tombs is hard, gritty work. The first step often is literally stumbling across bones that have eroded from a hillside. Then, there's some detective work tracking those fragments uphill to their source. There's the backbreaking work of moving tons of rock to expose layers that hold the ancient bones, followed by the painstaking excavation of sometimes fragile remains that haven't seen the light of day for millions of years.

Most of the time, it's achingly monotonous. But oh, those moments of excitement!

Sunday, July 2: Members of the dig team gather at noon at a hotel in Billings, Mont. Many stayed elsewhere the night before—some at hotels, others at campsites, a few at their nearby homes.

Our 11-vehicle caravan reaches the dig site, about 160 kilometers north of Billings, in a little more than 2 hours. The highways and gravel roads that we follow pass through a variety of landscapes, including ranchland dotted with small oil wells and sparse forests.

We pull into our campsite and pile out of the cars into rolling pastureland. All eyes are immediately drawn to a grim, gray scar on the other side of a small valley, a quarry where Murphy and other paleontologists have, on and off during the past couple of years, spent time unearthing the remains of two large dinosaurs. We are tempted to rush over there, but there is a campsite to set up.

Dave Hein, owner of the ranch, has mowed an area where we can pitch the cook tent and park the supply trailers. Portable toilets are towed to the far side of the campsite, and the camp showers are assembled next to the water truck—compared with digs in more-remote locations, this expedition will be posh, I am told. I will be privileged enough to sleep in the back of a truck.

While the other campers set up their tents, I chat with Hein to find out more about the site. Some of his wife's ancestors—five brothers from England, from whom Hein's 5E Ranch gets its name—settled here about a century ago. In 1985, Hein first found chips of fossilized bone lying on a hillside.

Then in 2003, he and his son used some earthmoving equipment at the site and came across a few large bones. Realizing the possible importance of the find, they turned to local experts. After a series of phone calls, Hein spoke to Murphy, who has since excavated bones at the site each summer.

Around the campfire after dinner, at Murphy's behest, we take turns introducing ourselves. Our group of 33 includes teenagers, retirees, museum volunteers, geologists, paleontologists, and even a theology professor. Only about half of us have been on digs before, and we are all itching to get our hands dirty.

We spend the rest of the evening in song, 2 hours of guitar- and coyote-accompanied ballads, folk tunes, and sing-along classics such as "Dead Skunk in the Middle of the Road."

Long after we stumble off to our sleeping bags, the coyotes are still singing.

Monday, July 3: We gather at the site to learn some basic digging skills. The standard-issue tools are awls—think ice picks on steroids—and stiff paintbrushes. We aren't to use the awls like ice picks, however—a motion that Murphy refers to as "Hitchcocking."

Instead, we are to gently pry apart layers of rock, brushing away the debris and inspecting our work zone regularly so that we won't damage any fossil before we realize it's there. Done right, it's slow going. Poke, pry, sweep, repeat. Fill up a gallon-size scoop with debris, and then dump it in a bucket. Six or eight scoops fill a bucket, and six or eight buckets fill a wheelbarrow. Roll the wheelbarrow downhill, empty it, return to your little section of strata. Fill, roll, empty, repeat. A ton of rock makes a pile much smaller than you'd think.

Late in the morning, the theropod tooth comes to light. No other bones of a meat eater have been found at this site, says Murphy.

In 2004, Murphy and his colleagues finished unearthing the bones that Hein had found, including four neck vertebrae, a portion of a femur, and almost a dozen ribs. They identified the dinosaur as a long-necked herbivore called a sauropod, and the team nicknamed it Ralph, after an earlier member of Hein's family whose homestead had been just a few hundred meters from where the Heins had found the dinosaur.

During the 2005 field season, the paleontologists unearthed seven more neck vertebrae and eventually uncovered Ralph's skull, upside down and half a meter away from the rest of him. Ralph's head had probably rotted off, rolled into that position, and then been buried by an ancient stream. Sauropod skulls are exceedingly rare, says Murphy.

The tooth tip that our team found may have broken off as a theropod fed on Ralph's carcass. The femur fragment found 2 years ago showed signs of having been gnawed on, Murphy notes.

After our lunch, further excavations near the tooth reveal a 20-cm-long fragment of another of Ralph's ribs. A few other, heavily eroded pieces of dinosaur bone turn up, but Murphy says that they probably aren't Ralph's because his fossil bones are usually in good condition.

We also come across plant fossils that may provide clues about the environment in which Ralph died. Cris E. Merta, a geologist from Sheboygan, Wis., notes that at first glance, the plants appear to have been similar to modern-day reeds, so the area may once have been a wetland.

The rocks that hold Ralph probably were deposited as sediment sometime between 150 million and 147 million years ago, says Melissa V. Connely, a geologist from Casper College in Wyoming.

Tuesday, July 4: After breakfast, Murphy takes a few of us over to a neighboring ranch. Last year's dig team found a few bones there beneath a light coating of sand, dirt, and bone chips. About 50 m away, the group discovered the end of a 2-m-long femur sticking out from the ground. Our job today is to remove the plaster jacket that has protected that bone during the winter—from harsh weather as well as from the sharp hooves of grazing cows. We'll then dig farther around the bone so that the fossil can be removed later in the week.

Because the rock is much harder here than it is at the Ralph quarry, we must learn the basics of using small air hammers driven by compressed air. We take turns shattering rock.

As we work several centimeters away from the femur, we're constantly on the lookout for previously undiscovered bones from the same dinosaur. We also have to be careful not to damage the crumbly end of the bone that had been exposed to the elements before its discovery.

The rock at this site breaks into pieces that are thumbnail-size or larger, so it's a challenge to brush them out of the way as we work. Nevertheless, all goes well until one of the team members doesn't lift his heel quite high enough as he steps backward across the femur. Whoops! Because much of the outer surface of the bone had been bonded together with liquid adhesives, a piece of that veneer the size of a legal pad sloughs off, taking a couple of handfuls of bone chips with it.

We stand frozen, mouths agape, and when we turn off the air hammer, the silence is deafening. Then, a series of quietly muttered curses. After taking a minute or so to recover a bit of composure, one of the team members sheepishly retrieves Murphy from the bone site nearby. We explain what happened, plead contrition, and brace ourselves for the worst. Obviously disappointed, Murphy stands mute for a few seconds and then says, "That's OK. If anyone ever tells you they've never broken a [dinosaur] bone, then they haven't really been digging."

Although this is the first truly sunny day of the week, we work under a cloud for the rest of the day.

Except for lunch: As a special treat, Hein has invited the dig team to join his extended family for a Fourth of July barbecue, replete with especially refreshing lemonade and watermelon. Someone breaks out a family photo album and shows us pictures of Ralph, the namesake of the sauropod we're excavating.

All too soon, though, we're back at the neighboring ranch, where we break up more rock. We come upon a few more fossils, which fortunately remain intact.

After dinner, around the campfire, we learn that discoveries by other team members at the Ralph quarry have slowed to a standstill. However, a few participants roaming the hills about 150 m away have found some fragments of bone, tracked them back to their hillside source, and dug out what appear to be a few tail vertebrae and fragments of a spike from a stegosaur.

Wednesday, July 5: Today, the center of attention shifts from Ralph to the new stegosaur. Further excavation yields more tail vertebrae as well as some limb and foot bones. This site, which stretches along the hillside no more than 6 m or so, is a flurry of pick-and-shovel activity. The small shelf that we've dug into the hillside can barely accommodate all the dig-team members who want to get in on the action.

I take pictures and do my best to stay out of the way. On the third day of the dig, the novelty of excavation has worn off and muscle aches have set in.

That evening, we're rewarded by exciting news. Parts of the stegosaur's tail vertebrae that we excavated weren't fused as they would be in an adult, so the creature may have been a juvenile, says Susannah Maidment, a paleontologist from the University of Cambridge in England.

Other features of the bones suggest that they represent *Hesperosaurus*—an exciting possibility, she notes, because only four other fossils of this stegosaur species have been discovered, all of which are in private collections. The one *Hesperosaurus* that has been described in a journal paper didn't include limb bones, such as the one we've found.

Thursday, July 6: Today at the cozy stegosaur quarry, my fellow diggers and I expose many new bones, including the end of a large one, possibly a femur, that seems to extend quite a distance into the hill.

At the dig site on the neighboring ranch, other team members uncover a tangle of bones that will have to be left for another expedition to excavate.

Despite the daylong efforts of the two team members who remain faithful to the Ralph site, no bones are forthcoming.

"You know what you did today?" Murphy asks them at dinnertime. "You just closed that quarry. You put Ralph to rest."

Friday, July 7: This morning is a frenzy of activity. We have only half a day in the field, during which we must extract some fossils, jacket others, break camp, and head back to civilization.

While some team members sketch the layout of the bones, the less artistic of us—myself included—work with compasses and measuring tapes. A team leader at each site assigns a code number to each bone or assembly, and someone records location data. Small, free-floating bones are wrapped in aluminum foil and labeled. Larger bones are tightly swathed in aluminum foil, then several layers of wet paper towels, then an outer coating of plaster-soaked burlap. Wrestling a fragile, several-hundred-pound lump of stone, bone, and plaster down a steep hill and into the back of a truck is challenging, to say the least.

Midafternoon, we head south to Billings. There, for the first time in a week, we can take a shower that is hot and lasts longer than 6 minutes. Then, we get together for dinner at a restaurant. At our final meal as a group, we each speak a few words about the experiences that we've had.

Murphy closes out the speeches by telling us how well our diverse group has listened, learned, and come together as a team.

He plans to immortalize Ralph this fall in a journal paper, giving him a scientific name that will distinguish him in academic circles as a new species. That article will represent a lot of hard work, Murphy notes, adding that we should all proudly consider ourselves a part of the dinosaur-discovery team.

Dinosaur Shocker!*

By Helen Fields
Smithsonian, May 2006

She is not your typical paleontologist. Neatly dressed in blue Capri pants and a sleeveless top, long hair flowing over her bare shoulders, Mary Schweitzer sits at a microscope in a dim lab, her face lit only by a glowing computer screen showing a network of thin, branching vessels. That's right, blood vessels. From a dinosaur. "Ho-ho-ho, I am excite-e-e-e-d," she chuckles. "I am, like, *really* excited."

After 68 million years in the ground, a *Tyrannosaurus rex* found in Montana was dug up, its leg bone was broken in pieces, and fragments were dissolved in acid in Schweitzer's laboratory at North Carolina State University in Raleigh. "Cool beans," she says, looking at the image on the screen.

It was big news indeed last year when Schweitzer announced she had discovered blood vessels and structures that looked like whole cells inside that *T. rex* bone—the first observation of its kind. The finding amazed colleagues, who had never imagined that even a trace of still-soft dinosaur tissue could survive. After all, as any textbook will tell you, when an animal dies, soft tissues such as blood vessels, muscle and skin decay and disappear over time, while hard tissues like bone may gradually acquire minerals from the environment and become fossils. Schweitzer, one of the first scientists to use the tools of modern cell biology to study dinosaurs, has upended the conventional wisdom by showing that some rock-hard fossils tens of millions of years old may have remnants of soft tissues hidden away in their interiors. "The reason it hasn't been discovered before is no right-thinking paleontologist would do what Mary did with her specimens. We don't go to all this effort to dig this stuff out of the ground to then destroy it in acid," says dinosaur paleontologist Thomas Holtz Jr., of the University of Maryland. "It's great science." The observations could shed new light on how dinosaurs evolved and how their muscles and blood vessels worked. And the new findings might help settle a long-running debate about whether dinosaurs were warm-blooded, cold-blooded—or both.

Meanwhile, Schweitzer's research has been hijacked by "young earth" creationists, who insist that dinosaur soft tissue couldn't possibly survive millions of years. They claim her discoveries support their belief, based on their interpretation of Genesis, that the earth is only a few thousand years old. Of course, it's not unusual for a paleontologist to differ with creationists. But when creationists misrepresent Schweitzer's data, she takes it personally: she describes herself as "a complete and total Christian." On a shelf in her office is a plaque bearing an Old Testament verse: "For I know the plans I have for you," declares the Lord, "plans to prosper you and not to harm you, plans to give you hope and a future."

It may be that Schweitzer's unorthodox approach to paleontology can be traced to her roundabout career path. Growing up in Helena, Montana, she went through a phase when, like many kids, she was fascinated by dinosaurs. In fact, at age 5 she announced she was going to be a paleontologist. But first she got a college degree in communicative disorders, married, had three children and briefly taught remedial biology to high schoolers. In 1989, a dozen years after she graduated from college, she sat in on a class at Montana State University taught by paleontologist Jack Horner, of the Museum of the Rockies, now an affiliate of the Smithsonian Institution. The lectures reignited her passion for dinosaurs. Soon after, she talked her way into a volunteer position in Horner's lab and began to pursue a doctorate in paleontology.

She initially thought she would study how the microscopic structure of dinosaur bones differs depending on how much the animal weighs. But then came the incident with the red spots.

In 1991, Schweitzer was trying to study thin slices of bones from a 65-million-year-old *T. rex*. She was having a hard time getting the slices to stick to a glass slide, so she sought help from a molecular biologist at the university. The biologist, Gayle Callis, happened to take the slides to a veterinary conference, where she set up the ancient samples for others to look at. One of the vets went up to Callis and said, "Do you know you have red blood cells in that bone?" Sure enough, under a microscope, it appeared that the bone was filled with red disks. Later, Schweitzer recalls, "I looked at this and I looked at this and I thought, this can't be. Red blood cells don't preserve."

Schweitzer showed the slide to Horner. "When she first found the red-blood-cell-looking structures, I said, Yep, that's what they look like," her mentor recalls. He thought it was possible they were red blood cells, but he gave her some advice: "Now see if you can find some evidence to show that that's not what they are."

What she found instead was evidence of heme in the bones—additional support for the idea that they were red blood cells. Heme is a part of hemoglobin, the protein that carries oxygen in the blood and gives red blood cells their color. "It got me real curious as to exceptional preservation," she says. If particles of that one dinosaur were able to hang around for 65 million years, maybe the textbooks were wrong about fossilization.

Schweitzer tends to be self-deprecating, claiming to be hopeless at computers, lab work and talking to strangers. But colleagues admire her, saying she's determined and hardworking and has mastered a number of complex laboratory techniques that are beyond the skills of most paleontologists. And asking unusual questions took a lot of nerve. "If you point her in a direction and say, don't go that way, she's the kind of person who'll say, Why?—and she goes and tests it herself," says Gregory Erickson, a paleobiologist at Florida State University. Schweitzer takes risks, says Karen Chin, a University of Colorado paleontologist. "It could be a big payoff or it could just be kind of a ho-hum research project."

In 2000, Bob Harmon, a field crew chief from the Museum of the Rockies, was eating his lunch in a remote Montana canyon when he looked up and saw a bone sticking out of a rock wall. That bone turned out to be part of what may be the best preserved *T. rex* in the world. Over the next three summers, workers chipped away at the dinosaur, gradually removing it from the cliff face. They called it B. rex in Harmon's honor and nicknamed it Bob. In 2001, they encased a section of the dinosaur and the surrounding dirt in plaster to protect it. The package weighed more than 2,000 pounds, which turned out to be just above their helicopter's capacity, so they split it in half. One of B. rex's leg bones was broken into two big pieces and several fragments—just what Schweitzer needed for her micro-scale explorations.

It turned out Bob had been misnamed. "It's a girl and she's pregnant," Schweitzer recalls telling her lab technician when she looked at the fragments. On the hollow inside surface of the femur, Schweitzer had found scraps of bone that gave a surprising amount of information about the dinosaur that made them. Bones may seem as steady as stone, but they're actually constantly in flux. Pregnant women use calcium from their bones to build the skeleton of a developing fetus. Before female birds start to lay eggs, they form a calcium-rich structure called medullary bone on the inside of their leg and other bones; they draw on it during the breeding season to make eggshells. Schweitzer had studied birds, so she knew about medullary bone, and that's what she figured she was seeing in that *T. rex* specimen.

Most paleontologists now agree that birds are the dinosaurs' closest living relatives. In fact, they say that birds *are* dinosaurs—colorful, incredibly diverse, cute little feathered dinosaurs. The theropod of the Jurassic forests lives on in the goldfinch visiting the backyard feeder, the toucans of the tropics and the ostriches loping across the African savanna.

To understand her dinosaur bone, Schweitzer turned to two of the most primitive living birds: ostriches and emus. In the summer of 2004, she asked several ostrich breeders for female bones. A farmer called, months later. "Y'all still need that lady ostrich?" The dead bird had been in the farmer's backhoe bucket for several days in the North Carolina heat. Schweitzer and two colleagues collected a leg from the fragrant carcass and drove it back to Raleigh.

As far as anyone can tell, Schweitzer was right: Bob the dinosaur really did have a store of medullary bone when she died. A paper published in *Science* last June presents microscope pictures of medullary bone from ostrich and emu side by side with dinosaur bone, showing near-identical features.

In the course of testing a B. rex bone fragment further, Schweitzer asked her lab technician, Jennifer Wittmeyer, to put it in weak acid, which slowly dissolves bone, including fossilized bone—but not soft tissues. One Friday night in January 2004, Wittmeyer was in the lab as usual. She took out a fossil chip that had been in the acid for three days and put it under the microscope to take a picture. "[The chip] was curved so much, I couldn't get it in focus," Wittmeyer recalls. She used forceps to flatten it. "My forceps kind of sunk into it, made a little indentation and it curled back up. I was like, stop it!" Finally, through her irritation, she realized what she had: a fragment of dinosaur soft tissue left behind when the mineral bone around it had dissolved. Suddenly Schweitzer and Wittmeyer were dealing with something no one else had ever seen. For a couple of weeks, Wittmeyer said, it was like Christmas every day.

In the lab, Wittmeyer now takes out a dish with six compartments, each holding a little brown dab of tissue in clear liquid, and puts it under the microscope lens. Inside each specimen is a fine network of almost-clear branching vessels—the tissue of a female *Tyrannosaurus rex* that strode through the forests 68 million years ago, preparing to lay eggs. Close up, the blood vessels from that *T. rex* and her ostrich cousins look remarkably alike. Inside the dinosaur vessels are things Schweitzer diplomatically calls "round microstructures" in the journal article, out of an abundance of scientific caution, but they are red and round, and she and other scientists suspect that they are red blood cells.

Of course, what everyone wants to know is whether DNA might be lurking in that tissue. Wittmeyer, from much experience with the press since the discovery, calls this "the awful question"—whether Schweitzer's work is paving the road to a real-life version of science fiction's *Jurassic Park*, where dinosaurs were regenerated from DNA preserved in amber. But DNA, which carries the genetic script for an animal, is a very fragile molecule. It's also ridiculously hard to study because it is so easily contaminated with modern biological material, such as microbes or skin cells, while buried or after being dug up. Instead, Schweitzer has been testing her dinosaur tissue samples for proteins, which are a bit hardier and more readily distinguished from contaminants. Specifically, she's been looking for collagen, elastin and hemoglobin. Collagen makes up much of the bone scaffolding, elastin is wrapped around blood vessels and hemoglobin carries oxygen inside red blood cells.

Because the chemical makeup of proteins changes through evolution, scientists can study protein sequences to learn more about how dinosaurs evolved. And because proteins do all the work in the body, studying them could someday help scientists understand dinosaur physiology—how their muscles and blood vessels worked, for example.

Proteins are much too tiny to pick out with a microscope. To look for them, Schweitzer uses antibodies, immune system molecules that recognize and bind to specific sections of proteins. Schweitzer and Wittmeyer have been using antibodies to chicken collagen, cow elastin and ostrich hemoglobin to search for similar molecules in the dinosaur tissue. At an October 2005 paleontology conference,

Schweitzer presented preliminary evidence that she has detected real dinosaur proteins in her specimens.

Further discoveries in the past year have shown that the discovery of soft tissue in B. rex wasn't just a fluke. Schweitzer and Wittmeyer have now found probable blood vessels, bone-building cells and connective tissue in another *T. rex*, in a theropod from Argentina and in a 300,000-year-old woolly mammoth fossil. Schweitzer's work is "showing us we really don't understand decay," Holtz says. "There's a lot of really basic stuff in nature that people just make assumptions about."

Young-Earth creationists also see Schweitzer's work as revolutionary, but in an entirely different way. They first seized upon Schweitzer's work after she wrote an article for the popular science magazine *Earth* in 1997 about possible red blood cells in her dinosaur specimens. *Creation* magazine claimed that Schweitzer's research was "powerful testimony against the whole idea of dinosaurs living millions of years ago. It speaks volumes for the Bible's account of a recent creation."

This drives Schweitzer crazy. Geologists have established that the Hell Creek Formation, where B. rex was found, is 68 million years old, and so are the bones buried in it. She's horrified that some Christians accuse her of hiding the true meaning of her data. "They treat you really bad," she says. "They twist your words and they manipulate your data." For her, science and religion represent two different ways of looking at the world; invoking the hand of God to explain natural phenomena breaks the rules of science. After all, she says, what God asks is faith, not evidence. "If you have all this evidence and proof positive that God exists, you don't need faith. I think he kind of designed it so that we'd never be able to prove his existence. And I think that's really cool."

By definition, there is a lot that scientists don't know, because the whole point of science is to explore the unknown. By being clear that scientists haven't explained everything, Schweitzer leaves room for other explanations. "I think that we're always wise to leave certain doors open," she says.

But Schweitzer's interest in the long-term preservation of molecules and cells does have an otherworldly dimension: she's collaborating with NASA scientists on the search for evidence of possible past life on Mars, Saturn's moon Titan, and other heavenly bodies. (Scientists announced this spring, for instance, that Saturn's tiny moon Enceladus appears to have liquid water, a probable precondition for life.) Astrobiology is one of the wackier branches of biology, dealing in life that might or might not exist and might or might not take any recognizable form. "For almost everybody who works on NASA stuff, they are just in hog heaven, working on astrobiology questions," Schweitzer says. Her NASA research involves using antibodies to probe for signs of life in unexpected places. "For me, it's the means to an end. I really want to know about my dinosaurs."

To that purpose, Schweitzer, with Wittmeyer, spends hours in front of microscopes in dark rooms. To a fourth-generation Montanan, even the relatively laid-back Raleigh area is a big city. She reminisces wistfully about scouting for field sites on horseback in Montana. "Paleontology by microscope is not that fun," she says. "I'd much rather be out tromping around."

"My eyeballs are just absolutely fried," Schweitzer says after hours of gazing through the microscope's eyepieces at glowing vessels and blobs. You could call it the price she pays for not being typical.

The Dino Wars[*]

Across the American West, Legal Battles Over Dinosaur Fossils Are on the Rise as Amateur Prospectors Make Major Finds

By Donovan Webster
Smithsonian, April 2009

Editor's note: On August 6, 2009, the 8th U.S. Circuit Court of Appeals upheld an earlier ruling that Ron Frithiof did not engage in fraud and that he and his team can retain ownership rights of Tinker the Tyrannosaurus. *For more on this story and other dinosaur-related news, read our Dinosaur Tracking blog.*

Buried beneath a barren stretch of South Dakota badland, the deceased appeared small for its species. As Ron Frithiof, an Austin, Texas, real-estate developer turned dinosaur prospector, dug cautiously around it in a rugged expanse of backcountry, he was growing increasingly confident that he and his partners were uncovering a once-in-a-lifetime find.

Ever since he had heard about a private collection going up for sale in the mid-1990s, Frithiof, now 61, had been hunting dinosaurs. "I'd thought fossils were things you could see only in museums," he says. "When I learned you could go out and find stuff like that, to keep or even to sell, it just lit a fire in my imagination. I studied every book I could, learned techniques of extraction. Fossils inspire a powerful curiosity."

Frithiof was keenly aware that the skeleton of a mature *Tyrannosaurus rex* ("Sue," named in honor of prospector Sue Hendrickson, who made the find in western South Dakota in 1990) had been auctioned off—at Sotheby's in New York City in 1997—for more than $8 million. The specimen that Frithiof and his fellow excavators began unearthing in 1998, in a painstaking, inch-by-inch dig was about four feet tall, less than half Sue's height. With unfused vertebrae and scrawny shin and ankle bones, the skeleton was almost certainly that of a juvenile. If so, it would likely be the most complete young *T. rex* ever discovered. A find of this magnitude,

Frithiof knew, would create a sensation. Its value would be, as he put it, "anyone's guess." $9 million? $10 million? This was uncharted territory.

For nearly three years, the excavators—including longtime fossil hunter Kim Hollrah, who had first investigated the site—continued their meticulous work. Whenever Frithiof, Hollrah and their companions could coordinate time off from work, they would drive 24 hours straight, from Texas to the dig site, north of Belle Fourche, South Dakota, which Frithiof had leased from a local rancher in 1998. "Most years, we'd spend about a month working," he recalls. "Thirty or 40 days a summer, before the weather would drive us off."

Braving blistering 100-degree temperatures, the crew took every precaution to keep the specimen intact. At the same time, they were attempting to wrest it from the ground before South Dakota's brutal winter set in. "That's one of the paradoxes of fossil collecting," says Frithiof. "Once a specimen is exposed to the elements, it's a race to get it out in as responsible a way as possible, to protect it from wind and rain and weathering. It's like a slow-motion race."

Paleontological excavation is nothing if not grueling. "We worked inch by inch, brushing bits of rock and soil away, taking a pin to strip away just that next little bit of rock and earth [to reveal the rough contours]," Frithiof told me. On a good day, an experienced fossil excavator might uncover only a few inches of skeleton. Frithiof and the others gingerly pried out each section, still enclosed in the crumbly chunk of rock matrix that had originally surrounded it. In preparation for transport, the prospectors then wrapped the sections in layers of tissue paper, aluminum foil and plaster.

As the dig moved forward, Frithiof's colleagues, with a nod to "Sue" (today a centerpiece attraction at Chicago's Field Museum), decided the new *T. rex* needed a name. The one they came up with honored Frithiof's role as the project's financial backer. "I don't know why my parents started calling me Tinker," says Frithiof. "Somehow, it stuck."

In 2001, as the excavation of Tinker headed toward completion, the team made another remarkable discovery: evidence of two additional *T. rex* skeletons on the site. By that point, a children's museum in the Midwest had indicated its willingness to pay up to $8.5 million for Tinker. During the prospective purchaser's pre-transaction research, however, a massive legal hiccup was uncovered—one that Frithiof and his lawyers would later insist had been an honest mistake.

Tinker, as it turned out, had been found not from local rancher Gary Gilbert's land but from adjacent property owned by Harding County, South Dakota. In November 2000, Frithiof, he says, with an eye to future excavations, had leased the parcel from the county; the agreement stipulated that the county would get 10 percent of the sale price for any fossils uncovered there. Now, in August 2004, Harding County filed a civil lawsuit in Federal District Court against Frithiof and his partners alleging fraud, trespass and conspiracy.

Frithiof's world caved in. After devoting years to Tinker, the prospector was suddenly in danger of going to jail for his efforts. "This whole experience has been a disaster," he says. "[With] all the lawyers' fees, not to mention the disruption of my life, it's cost me a fortune. And it's been very hard on my family. You gotta remember, I've never been in trouble in my life. Not even a traffic ticket." The disputed dinosaur, according to Frithiof's attorney Joe Ellingson, "wrecked my client's life."

Moreover, the fossil was consigned to limbo. As a result of byzantine twists in the litigation, Tinker's bones would soon be placed under another lawyer's supervision, stored in plastic tubs at an undisclosed location in Harrisburg, Pennsylvania—1,400 miles from the excavation site.

Across the American West and Great Plains, an intensifying conflict over the excavation of fossils—everything from a five-inch shark's tooth, which might sell for $50, to Frithiof's spectacular *T. rex*—has pitted amateur excavators against both the federal government and scientists. Scores, perhaps thousands, of prospectors—some operating as poachers on federally protected land—are conducting digs across hundreds of thousands of square miles from the Dakotas to Texas, Utah, Wyoming and Montana.

"In terms of digging for fossils, there are a lot more people" than there used to be, says Matthew Carrano, curator of dinosauria at the Smithsonian Museum of Natural History. "Twenty years ago, if you ran into a private or commercial fossil prospector in the field, it was one person or a couple of people. Now, you go to good fossil locations in, say, Wyoming, and you find quarrying operations with maybe 20 people working, and doing a professional job of excavating fossils."

Fueling the frenzy is skyrocketing market demand, as fossils, long relegated to the dusty realm of museum shelves, have entered the glitzy spheres of home décor and art. "There have always been private fossil collectors," says David Herskowitz of Heritage Auction Galleries in Dallas. "The difference is, historically, a private fossil collector was wealthy. But today, interest in fossils has grabbed the attention of a broad swath of the population. That means a lot more people are collecting."

Who's buying these days? Just about anyone. With prices to suit virtually any budget, one can own an ancient remnant of life on earth: a botanical fossil, such as a fern, may cost as little as $20; a fossil snail, perhaps, may well go for $400.

The real action, however, is in the big vertebrates: dinosaurs that roamed the earth between 65 million and 220 million years ago. These are the specimens attracting the high rollers—serious collectors. Actors Harrison Ford and Nicolas Cage, for example, are rumored to have impressive collections.

The paleo-passion, however, extends far beyond celebrities. "The group who used to be serious fossil collectors—that's really grown," says money manager Charles Lieberman of Advisors Capital Management in Hasbrouck Heights, New Jersey. At his office, Lieberman displays several impressive specimens, including a three-foot-long Cretaceous herbivore, *Psittacosaurus*. "Since the book and movie *Jurassic Park*," he adds, "interest in fossil collecting has gone into overdrive, affecting demand and elevating prices."

The rise in prices is fueling the prospecting boom in the Great Plains and West— not necessarily because of a higher concentration of fossils there, but because the American West is one of the world's easiest places to find them. "If you had flown around the world 150 million years ago, the West wouldn't be more populated by dinosaurs than anywhere else," says the Smithsonian's Carrano. "But in the West, the rock layers laid down during the age of dinosaurs are currently exposed. It also helps that the landscape is dry, so there's not a lot of vegetation covering the rock. And it's erosive, so new rock is constantly being uncovered."

While fossils can now be found in stores from Moab to Manhattan, the most unusual (and valuable) specimens tend to show up at auction houses—or vanish into the shadowy world of private purchasers, some of whom are buying on the black market. At the Tucson Gem and Mineral Show, for instance, it is possible to obtain illegally taken fossils. While Carrano does not attend the show, it's well-known, he says, that, "if you spend the week building trust with some of the sellers, you'll get invited back to a hotel room and be shown exquisite fossil specimens that were probably taken illegally. We're talking museum-grade specimens that are going to disappear into private collections."

The auction houses, of course, make sure their offerings come with documented provenance. In only a few hours in April 2007, Christie's in Paris gaveled off fossils worth more than $1.5 million—including a dinosaur egg that went for $97,500 and the fossilized skeleton of a Siberian mammoth that fetched $421,200. In December 2007, a 70-million-year-old mosasaur—a 30-foot carnivorous underwater reptile excavated in North Africa—brought more than $350,000 at Los Angeles auctioneer Bonhams & Butterfields. In January 2008, Heritage Auction Galleries in Dallas sold the largest mastodon skull ever found for $191,000 and a 55-million-year-old lizard from the Dominican Republic, its flesh and skin preserved in amber, for $97,000. "The day's tally was $4.187 million," says auction director Herskowitz. "While I can't disclose who my buyers were, I can say many of them have small to substantive museums on their properties."

Then there's eBay. When I logged on recently, I discovered 838 fossil specimens for sale, including a spectacular ammonite—an ancestor of today's chambered nautilus—expected to go for upward of $3,000. Very little was disclosed about where any of the fossils came from. "Here's what I can tell you about eBay," says Carrano. "If a fossil being sold there comes from Morocco, China, Mongolia, Argentina or a number of other nations, at some point it was part of an illegal process, since those countries don't allow commercial fossil export."

In the United States, the law regulating fossil excavation and export is far from straightforward. Property statutes state that any fossil taken with permission from privately owned land may be owned and sold—which is why legitimate excavators usually harvest fossils from individual landowners. A complex series of regulations apply to fossils removed from federal and state land (including Bureau of Land Management [BLM] tracts, national forests and grasslands, and state and national parks) and what are known as jurisdictional lands—for example, the public land held by Harding County, South Dakota.

To complicate matters, some fossil materials—limited amounts of petrified wood or fossil plants, for example—may be removed from certain public lands without oversight or approval. In most cases, however, permits are required; applications are reviewed according to a time-consuming process. Prospectors who want to cash in quickly on a single find are often reluctant to abide by the law. Given that there are nearly 500 million acres of publicly held land in the United States (two-thirds of which contain some of the best excavation zones in the world), prospectors who dig illegally are not often caught. "Newly harvested fossils are flooding the commercial market," says Larry Shackelford, a special agent with the BLM in Salt Lake City. "Running down each one and checking where it came from? We don't have the manpower."

In fact, law enforcement officials can barely keep up with prosecutions already underway. Although state and federal officials may not discuss cases currently in litigation, they acknowledge that volume is increasing. "In most districts, we easily see one or two new leads a month," says Bart Fitzgerald, a BLM special agent in Arizona. "Mostly these become civil cases. We understand that enthusiasm gets the best of people sometimes. Someone finds an amazing fossil and they take it home. Mostly we just want to recover the fossil—it's government property. But once in a while, we see a case where clearly the intent was criminal: where people were knowingly extracting fossils from public land for private profit. Those we prosecute criminally."

A major criminal case began unfolding in 2006, when a largely intact *Allosaurus*—a meat-eating older cousin of *T. rex*—was taken from public land in Utah. The excavator went to great lengths to look legitimate, including creating bogus letters of provenance. The dinosaur bones were first transported from Utah to a U.S. buyer, then to a purchaser in Europe, before finally being sold to a collector in Asia. In February 2007, the *Allosaurus* poacher—who had been turned in anonymously—was convicted on one count of theft of federal property.

Several years earlier, a high-profile case involved paleo-prospector Larry Walker, who discovered a cache of fossil *Therizinosaurs*—a rare dinosaur/bird hybrid—in the desert outside his Moab, Utah, hometown. Working at night beneath camouflage netting, Walker excavated 30 to 40 of the creatures' distinctive ripping claws, then sold the specimens at the Tucson Gem and Mineral Show for a total take of roughly $15,000.

"He knew what he was doing was illegal," says Loren Good, a special agent for the BLM's Idaho district. "Working with the FBI, we did a joint investigation into the source of the claws and prosecuted Mr. Walker. He received a ten-month incarceration and a $15,000 fine."

"These cases come in all forms," says the BLM's Fitzgerald. "Take the example of some tour operators in Montana. They took a group of tourists out recently on a fossil-hunting trip, strayed onto public land and extracted fossils from a good site there. Was it an honest mistake or a calculated commercial move?" Fitzgerald asks. "After all, the tour operators carried GPS units; they knew precisely where they were." (Charges have not yet been filed.)

In the Tinker case, the prosecution claimed that Frithiof knew he was on county property when he found the Tinker specimen, that he had signed the agreement with Harding County without informing officials of the find and that he had negotiated a perhaps $8.5 million sale without telling the county. "Harding County believes Mr. Frithiof first discovered the specimen's location, then induced the county into a lease, knowing the value of what existed on the property without disclosing it to us," says Ken Barker, a Belle Fourche, South Dakota, attorney retained by the county to prosecute the case. "Because of this, we seek to void the lease agreement, entered into fraudulently, and to recover the county's property."

Frithiof sees things differently. It wasn't until the prospective purchaser's survey in 2001, he says, that all parties learned that the Tinker site was on county land. "We were something like 100 feet across the [county] property boundary," he says. "Even the rancher we were working with believed we were on his land. It was an honest mistake. And I already had a lease on that land with Harding County.

"It wasn't like we were sneaking around," Frithiof adds. "Our find had been in the newspaper. We'd been on the Discovery Channel. We'd had prominent paleontologists, like Bob Bakker from the University of Colorado, out to look at it. What we were doing was all out in the open. Nobody thought we were doing anything illegal . . . at all."

In June 2006, Judge Richard Battey of the United States District Court voided the agreement between Frithiof and the county and ruled, on the basis of a technicality, that Tinker belonged to Harding County. Frithiof appealed. In September 2007, a United States Court of Appeals panel reversed the decision. The Tinker fossil, they ruled, was Frithiof's property; only the original contract's 10 percent payment was owed to Harding County. The appeals court then sent the case back to Federal District Court for final disposition. Frithiof had no choice but to wait.

In the meantime, the location of Tinker—and the fossil's condition—had become a source of contention. Before the legal wrangling began, Frithiof had delivered sections of the skeleton to private curators Barry and April James, who specialized in preparation of paleontological specimens for display, at their Sunbury, Pennsylvania, firm, Prehistoric Journeys. (The process involves removal of the stone matrix encasing the excavated bones.) Once the litigation proceeded, however, the Jameses, who say they had put $200,000 worth of labor and more than two years into the project, were barred from completing the work or collecting payment from Frithiof. Their company filed for bankruptcy in 2005.

"Now I have the Tinker fossil in my possession," says Larry Frank, a Harrisburg, Pennsylvania, attorney who is trustee of the James bankruptcy. "I've filed an artisans' lien against the value of the specimen. Until the matter is resolved, the skeleton will sit in large plastic containers in my possession. We believe that's a good, safe place for it."

For scientists, commercial excavation of fossils—legal or not—raises troubling questions. "For me," says Mark Norell, chairman and curator of vertebrate paleontology at the American Museum of Natural History in New York City, "the big concern with all this private digging is that it may be robbing science of valuable knowledge."

Norell believes that anyone harvesting fossils "needs to be considerate of scientific data surrounding the specimen." Context is important. "A lot of the guys out there digging commercially are just cowboys; they don't care about the site where the fossil sits, how it's oriented in the earth, what can be found around it to give us clues to what the world was like when that fossil animal died." Some commercial excavators "want only to get the specimen out of the ground and get paid—so we lose the context of the site as well as the fossil itself."

The Smithsonian's Carrano says all scientifically significant fossil specimens, whether from public or private lands, should be placed into museums for study in perpetuity. "Any unique fossil has more value scientifically and educationally than we can ever place a cash value on," he adds. "In a perfect world, there'd be a way to vet every fossil collected: the significant ones would be retained and studied; others could go to commercial use. Not every fossil shark's tooth is significant, but some are. Let's retain those significant ones for study."

For the past several years, the Society of Vertebrate Paleontology, one of the fossil world's preeminent professional organizations, has lobbied in support of Congressional legislation that would protect fossils taken from public lands. Since 2001, a bill introduced by Representative James McGovern, Democrat of Massachusetts—the Paleontological Resources Preservation Act—has languished in both the House and Senate. The delay, some proponents believe, stems from some western lawmakers' reluctance to add any regulations regarding public lands. If passed into law, the act would require that only trained, federally certified professionals be allowed to extract fossils from public lands—and would substantially increase penalties for illegal fossil excavation.

The proposed legislation has galvanized critics, from mining company executives to paleontology prospectors, many of whom argue that improved enforcement of existing laws is all that is needed. "This new bill provides no funding for additional federal agents to police these areas, meaning it has no teeth," says Jack Kallmeyer, a paleontological prospector. "As long as there is demand for the commodity, without sufficient enforcement personnel, nothing will stop illegal collecting."

Kallmeyer also notes that proposed and existing fossil-extraction laws do not address a critical threat to the nation's fossil heritage. "There are a number of dinosaur and [other] vertebrate fossils out there [on public lands] that are not rare. Professional paleontologists aren't interested in excavating them, as those specimens are well known and well studied. Why shouldn't amateur or commercial collectors be allowed to extract those?" Fossils left exposed over years, Kallmeyer adds, will eventually erode away.

But paleontologist James Clark of George Washington University in Washington, D.C., who serves on the government liaison committee for the Society of Vertebrate Paleontology, disagrees. "Nobody knows how much fossil material is being taken off public lands and smuggled out," he says. "We don't know the scale of what's being lost." Clark, who sees the proposed federal bill as a step forward, believes that existing legislation is too nonspecific and confusing. "As it stands now, the situation is a free-for-all," he says.

Through the winter of 2007–2008, as Frithiof awaited another ruling from Federal District Court, he and lawyer Joe Ellingson hunkered down. "We don't want to say much," Ellingson told me. "We don't want to antagonize anyone in any way. We just want to wait and get our ruling."

The delay, however, proved excruciating for Frithiof, who continued living near Austin, selling real estate. "There's not one hour," he says, "that it wasn't in the back of my mind. And that takes a toll. Even a physical toll." Frithiof says he developed cardiac problems. "I just want this all to be over," he says, "so I can go back to my site and keep working. We've found evidence of two other *T. rex* specimens there, but we don't know if they're complete or not. We've covered them up to protect against the elements. Until all this is resolved, we've been barred from working."

At last, on February 5, 2008, Judge Battey ruled that Frithiof's lease with Harding County was legal and enforceable. Frithiof owned Tinker, though he would have to give the county 10 percent of any profits from its sale. Harding County, the decree said, "knowingly entered into this contract, and now must live with the consequences of its actions." For Frithiof, the ruling meant "a huge weight had disappeared off my life."

But within weeks of the ruling, Harding County appealed yet again, sending the case back into court and consigning Frithiof once again to legal limbo. After more than four years of litigation, disposition of the appeal is expected within weeks. "This experience has removed the joy of fossil hunting for me," says Frithiof. "I haven't done one day of digging since the day initial charges were brought."

And yet, Frithiof tells me, an even larger question preoccupies him. "My thoughts always return to the exposed fossils out there on our public land," he adds. "Fossils that are going unexcavated due to lack of interest. The ones paleontologists are never going to extract because they are fossils that are too common, but which some collector might cherish."

Frithiof insists that careful amateur excavators can make a significant contribution to science. "The fossils are out there, wind and rain weathering them, while people argue about who is allowed to collect them and who isn't. After a year or two of exposure, any fossil begins to disintegrate and crumble to dust." And then, he adds, "Well, nobody gets them. They're just gone."

3

Of a Feather? The Bird-Dinosaur Link

A specimen of *Microraptor gui* on display at the Paleozoological Museum of China, in Beijing.

A feathered *Deinonychus antirrhopus* model at the Royal Ontario Museum, Toronto, Canada.

Editor's Introduction

Today it is commonly accepted that birds and dinosaurs share evolutionary ties, but from feathers and fingers to bones and wings, there are many examples of missing links scientists have yet to discover. The articles in this chapter focus on some of these, exploring the bird-dinosaur connection.

Carolyn Gramling, a science reporter for *Earth* magazine, opens the chapter with "Are Birds Dinosaurs? New Evidence Muddies the Picture." Starting with the Darwin-era discovery of the prehistoric bird *Archaeopteryx* and presenting evidence collected through the modern day, Gramling outlines the current debate surrounding the relationship between dinosaurs and birds. The evidence can sometimes be conflicting, and as Gramling reports, many types of dinosaurs appear to have had feathers—or feather-like "dino-fuzz"—but not bird-like respiratory systems.

In "Which Came First, the Feather or the Bird?" the next article, Yale ornithologist Richard O. Prum and University of Connecticut Emeritus Professor Alan H. Brush explain how their research and findings by colleagues explain the evolution of modern bird feathers. The emerging field of evolutionary development biology, or "evo-devo," has converged with recent fossil discoveries to suggest theropod dinosaurs, the suborder that included *T. rex*, sprouted feathers "before the origin of birds or the origin of flight," the authors write.

In the next piece, "Bird's-Eye View," Smithsonian curator of dinosauria Matthew T. Carrano and Ohio University anatomy professor Patrick M. O'Connor use cladistics, an area of biology concerned with the evolutionary links between organisms, to compare dinosaurs with modern birds. The authors discuss how the two are related and explain how it's possible to use what we know about birds to devise theories about dinosaur size, locomotion, and bone characteristics.

From there, *New Scientist* writer Jeff Hecht takes us to China, a veritable "Dinotopia," given its recent spate of stunning fossil finds. In Liaoning province there is a rock formation called Yixian that boasts some of the greatest fossil diversity in the world. There, researchers have found not one, but several species of feathered dinosaurs that could fill the missing link between modern-day birds.

Additional evidence for a bird-dinosaur connection comes from an unexpected source in "Origin of Species: How a *T-Rex* Femur Sparked a Scientific Smackdown," the chapter's final selection. Writing for *Wired* magazine, Evan Ratliff reports on a scientific debate in bioinformatics—a field that involves using computers to analyze bones and protein—pertaining to portions of a *T. Rex* femur found to share commonalities with birds. The standards and methods are complex, but Ratliff does a good job of explaining the benefits and perils of this new approach.

Are Birds Dinosaurs?*

New Evidence Muddies the Picture

By Carolyn Gramling
Earth, October 2009

In 1861, German paleontologist Christian Erich Hermann von Mayer excavated the fine-grained limestone layers of a quarry near Solnhofen, Germany. The 150-million-year-old limestone had already proven promising for finding fossils: A year earlier, von Mayer had found the imprint of a single feather preserved in the rock. But this time, he discovered something more spectacular: an entire skeleton of what appeared to be an ancient bird.

Charles Darwin described the creature, which was missing its head and neck, in the fourth edition (1866) of his book "On the Origin of Species": "Until quite recently . . . some have maintained that the whole class of birds came suddenly into existence during the Eocene period . . . but now we know . . . that a bird certainly lived during the deposition of the upper greensand," a term used at the time to describe a Cretaceous-aged green sandstone formation. Darwin went on: "That strange bird, the *Archaeopteryx*, with a long lizard-like tail, bearing a pair of feathers on each joint, and with its wings furnished with two free claws, has been discovered in the oolitic slates of Solnhofen."

Following von Mayer's discovery, scientists began to consider that dinosaurs—specifically theropods, bipedal dinosaurs that evolved during the Late Triassic about 230 million years ago—might be the ancestors of birds. Theropods, the group of dinosaurs that includes velociraptors and tyrannosaurs, had certain features in common with modern birds: three toes, a wishbone and air-filled spaces in their bones. And paleontologists continue to unearth fossil and morphological data—including hundreds of skeletal features, similarities in digestive systems and other internal systems, and possible feathered dinosaur fossils—that support this hypothesis.

One such powerful link came in 2005, when the details of a nearly complete skeleton of another Solnhofen-originating *Archaeopteryx* were published in *Science*. The new specimen was the best-preserved yet, revealing that *Archaeopteryx* lacked a modern bird's first toe—the one that a bird uses to perch—but that it did have a hyperextendable second toe, similar to the menacing claw of a velociraptor. But many evolutionary puzzles remain: How did hands become wings? Did any dinosaurs have feathers? How did birds' unusual, highly flight-specific respiratory systems evolve? Paleontological finds provide evidence both for and against dinosaurian ancestry—and each new dinosaur discovery only seems to further muddy the picture.

BIRDS OF A FEATHER?

In 1996, scientists in China discovered *Sinosauropteryx*, an Early Cretaceous dinosaur from about 120 million years ago that bore a resemblance to the small, voracious "compys" (*Compsognathus*) portrayed in the movie "Jurassic Park." The *Sinosauropteryx* fossils were remarkably well-preserved and appeared to show that the creature's body was covered with faint, furry protofeathers, dubbed "dino-fuzz."

China proved to have many more small, bipedal feathered dinosaurs dating to the Early Cretaceous: *Sinornithosaurus millenii*, discovered in 1999, showed traces of avian feather structures. In 2002, scientists announced the discovery of a dinosaur, named *Microraptor gui*, that lived about 125 million years ago and appeared to have feathers on both its front and hind limbs. *Beipiaosaurus*, a 125-million-year-old dinosaur discovery announced in January, was the first to show feather-like structures that had a single filament rather than multiple filaments—possibly showing the earliest example yet of feather evolution.

These discoveries seemed to provide a rough sketch of feather evolution in these theropod dinosaurs, supporting the idea that theropods might be the ancestors of today's birds. Then, last March, scientists announced yet another small, bipedal Early Cretaceous dinosaur that may have had protofeathers, called *Tianyulong confuciusi*. But this time, there was a twist. Unlike the other discoveries, *Tianyulong confuciusi* was not a "lizard-hipped" theropod; instead, the new dinosaur was an ornithischian, described as "bird-hipped," with a pelvis in which the pubis points backward (for example, *Stegosaurus* and *Triceratops* were both bird-hipped).

The terminology is misleading: Although this hip structure is somewhat similar to a bird's, paleontologists have not considered the hip structure to be evidence of ancestry. Instead, ornithischians and birds likely evolved the backward-facing hip structure separately, in an example of convergent evolution. But the new evidence of dino-fuzz on a bird-hipped dinosaur ruffled feathers, suggesting that an ancestral relationship between dinosaurs and birds may be more complex than paleontologists had realized.

Feathered dinosaurs aren't the only recent evidence that they may be birds' ancestors; scientists have also found that some theropods may show traces of birds' complex breathing structure. In September 2008, paleontologist David Varricchio of Montana State University in Bozeman reported in PLoS ONE that the skeleton of a large predatory dinosaur held pockets that may have contained bird-like air sacs. The dinosaur, called *Aerosteon riocoloradensis*, showed signs of the air sacs in its wishbone and ribs. And other researchers, Varricchio says, have found evidence of air sacs in other theropods.

"In our paper, we tried to take the approach that it's a complex system in modern birds," Varricchio says. "Not all dinosaurs [had] that system; but within theropods, as you move closer to birds, you get additions of more and more parts of the machine." For example, the earliest theropods had only the cervical air sac, he says, but later theropods show signs of cranial, clavicular and abdominal air sacs as well. Another important aspect of birds' breathing is the sternum and the ribs, which "act as a bellows-pump mechanism," pushing air under pressure along a one-way path through the bird's rigid lungs, Varricchio says. It's a structure, he says, that only shows up in theropods that are close to the ancestry of birds.

Perhaps the most compelling—and controversial—evidence of the close ties between birds and theropods is found at the molecular level. In 2005, paleontologist Mary Schweitzer of North Carolina State University in Raleigh and colleagues described how they extracted soft tissue from a 68-million-year-old *Tyrannosaurus rex* fossilized leg bone. By 2007, Schweitzer and her team identified small fragments of collagen in the tissue. They compared protein sequences in the collagen with those of modern animals—and found the closest match to be modern chickens, strengthening the genetic link between dinosaurs and birds.

OR NOT DINOSAURS AFTER ALL?

But for every argument, it seems there's a counterargument. Not everyone thinks dinosaurs deserve a place in the bird family tree. Ornithologist Alan Feduccia of the University of North Carolina at Chapel Hill, for example, doubts the dino-fuzz: He sees skin and tissue where paleontologists see feathers. The numerous fossil imprints of protofeather filaments in *Sinosauropteryx*, for example, represent fibers of collagen, the primary structural protein in skin tissue, Feduccia and others wrote in a *Journal of Morphology* study published in 2005. Collagen is relatively tough, composed of inelastic fibers and relatively insoluble in water, making it more likely than other soft tissues to be preserved. Another strike against the idea that *Sinosauropteryx* represented a stage in feather evolution is that *Archaeopteryx*, about 30 million years older, could already fly, according to Feduccia.

Meanwhile, last June, zoologist Devon Quick and vertebrate paleobiologist John Ruben of Oregon State University in Corvallis published a paper in the *Journal of Morphology* that emphasizes the uniqueness of bird respiration—specifically the skeletal morphology related to avian breathing—and on that basis calls into question a theropod-bird ancestry.

A bird's respiratory system stands in stark contrast to that of other animals, Quick says. Flying requires a lot of energy and oxygen. As a result, over time, birds have developed a highly efficient lung and respiratory system that allows them to take in enough oxygen and exchange carbon dioxide efficiently enough to allow them to fly. "The way they move air across their lungs is really different from the way we do it," she says. "It's very special. We use a diaphragm to change the volume of our lungs. They don't change the volume at all, because they have these really specialized collapsible structures, really thin-walled, compliant air sacs." Birds also have a special skeleton, Quick says, that keeps the air sac from collapsing when the bird inhales.

In the study, Quick and Ruben detailed new findings about this specialized skeleton, including an immobile thigh bone that is locked into the body wall and provides extra skeletal support for the birds' flabby air sac. But the more controversial news was the underlying implication: that if dinosaurs are the ancestors of birds, it seems unlikely that scientists would have found no trace of this highly specialized system in any dinosaur fossils.

"We're suggesting that theropod dinosaurs did *not* have a bird-like lung," Quick says. That, in turn, suggests that theropods may not be the ancestors of birds, she says, but instead may represent an extinct lineage. "I don't think it's clear what theropods are at all, as far as what they gave rise to or what gave rise to them." One possibility, she says, is that theropods and birds might both be derived from a common ancestor.

The evolution of avian respiration is a tricky question, Varricchio says, because it's a complex system that's not likely to be preserved in the fossil record. "We're really trying to predict what the lungs [of dinosaurs] look like, and they don't leave any trace on any bone." Even in birds, it's not necessarily straightforward, he adds: For example, birds brooding on a clutch of eggs can't use their sternums as a bellows—they have to use their abdominal muscles, suggesting that although the sternum is important, it may not be absolutely necessary for birds to breathe. Still, in their paper, Quick and Ruben do make a good point that the abdominal structure of most theropod dinosaurs is distinct from that of birds, Varricchio says. So, he adds, that structure is "probably not doing the exact same thing" when it comes to breathing.

Quick insists that she didn't set out to disprove a bird-dinosaur link. "I just don't think we have enough information to make a definitive conclusion," she adds. "[It's] the nature of the fossil record—it's so spotty. While we can have some good preservation, we don't have the whole picture to say that 'x' is really derived from 'y.'"

A BIRD IN THE HAND OR NOT?

Quite apart from questions of feathers and lungs is another puzzle: how a theropod's hand might have evolved into a bird's wing. The hand-to-wing transition is a big thorn in the side of the theropodbird evolutionary relationship. The longstanding debate centers on digits: Although early theropods had five digits, they appeared to lose two of them over time, so that later theropods had only three. Modern birds also have three "fingers" in their wings.

So far, so good. The sticking point, however, is that although both theropods and birds had three digits, they don't seem to be the same three digits. Later theropods are missing the two outermost digits, or numbers four and five (equivalent in humans to the ring and pinky fingers). However, evidence from the embryos of developing birds suggests that their wings arise from digits two, three and four (so that they are, essentially, missing digits one and five—equivalent to the human thumb and the pinky).

That offset of digits has been a big enough problem that creationists have pointed to the issue as a way of discrediting evolutionary theory, says paleontologist James Clark of George Washington University in Washington, D.C. "More rationally," he adds, "the developmental biologists have looked to say, 'What's really going on here?'" But to date, the possible explanations for why theropods and birds each have three fingers—but not the same three fingers—have lacked supporting evidence. One idea, Clark says, was a "frame shift," in that the three bird fingers took on the shape of the three theropod fingers. "There was a lot of speculation about what happened," he says, "but there wasn't evidence from the fossils."

But an intriguing new fossil may fill in some gaps. Clark and colleagues published a paper in *Nature* last June describing a 160-million-year-old dinosaur found in a Jurassic mud pit in China. "These animals had become mired in the mud—there's evidence that they moved the sediment around and were struggling," Clark says. He and his team discovered a number of fossils in this mud pit, he says—it became the subject of a National Geographic film titled "Dino Death Trap"—"but we hadn't gotten a really good one, with a skull, until 2005."

That year, the team found a nearly complete skeleton that was missing only the last half of its tail. "Once we saw the skull and arms of this thing, we were pretty sure it was something new," Clark says. The creature, which they named *Limusaurus inextricabilis* ("mire lizard who could not escape"), had an unusual skull, with no teeth and a fully developed beak. It also had gizzard stones—something found in plant-eating chickens and turkeys, but not in carnivorous birds.

But the most interesting part of the new dinosaur was its hands: *Limusaurus* had four fingers, not three—and that, Clark says, suggests that it might represent a transitional theropod species. Next to its fully developed second, third and fourth fingers, *Limusaurus* had a tiny first digit. "It was reduced to a little nubbin, while the second digit had become this big robust finger," Clark says.

The wrist bones, meanwhile, also indicate some complicated developmental shifts, Clark says. *Limusaurus'* wrist bones were similar to bones associated with digits two, three and four, but the finger bones themselves resembled digits one, two and three. Overall, he says, the new dinosaur suggests that something transitional was occurring between the early theropods with five fingers and the later three-fingered theropods, such as allosaurs, which some scientists think gave rise to birds.

But not all paleontologists—even those who argue that birds are the descendants of theropod dinosaurs—are convinced that *Limusaurus* represents a missing link in hand-wing evolution, or that finding such a missing link is even necessary to prove the point. In fact, says Kevin Padian, a paleontologist at the University of California at Berkeley, the whole question of wing digits may be something of a straw man argument.

"Developmental biologists really don't know what they're looking at when they see the digits develop in a bird," Padian says. "They just put arbitrary numbers on them. For some reason, the authors think they have to reconcile an imaginary model that the bird fingers are two, three and four with the knowledge that as dinosaurs reduced their fingers, they lost four and five and wound up with one, two and three, which is what birds retain." As for *Limusaurus*, Padian says, "this seems to be just an ordinary theropod with some unusually reduced digits. That's not odd—lots of animals have weird fingers. But we don't use the weird ones to try to overthrow everything we know about how the fingers became reduced in dinosaurs."

So, are birds modern dinosaurs? The debate still continues. Ornithologist Feduccia's opposition to a dinosaurian ancestry for birds has become well-known. Some of his concern, he says, is that the link is so tenuous that it invites opposition from creationists, who point to any holes in the evolutionary link between birds and dinosaurs or disputes over evidence that dinosaurs were feathered as evidence against evolution. "We all agree that birds and dinosaurs had some reptilian ancestors in common," Feduccia told the University of North Carolina's Chapel Hill news service in 2005. "But to say dinosaurs were the ancestors of the modern birds we see flying around outside today because we would like them to be is a big mistake." Varricchio, however, frames the question differently. "Sometimes people simplify things and say that if birds do it, then dinosaurs must do it," he says.

"But dinosaurs are really diverse—they lived for 165 million years, not counting birds. They had different body shapes and likely did all kinds of different things. There's great variety and no reason a *T. rex* or another dinosaur should necessarily have the same apparatus. But that doesn't mean they couldn't give rise to birds."

Which Came First, the Feather or the Bird?*

By Richard O. Prum and Alan H. Brush
Scientific American, March 2003

Hair, scales, fur, feathers. Of all the body coverings nature has designed, feathers are the most various and the most mysterious. How did these incredibly strong, wonderfully lightweight, amazingly intricate appendages evolve? Where did they come from? Only in the past five years have we begun to answer this question. Several lines of research have recently converged on a remarkable conclusion: the feather evolved in dinosaurs before the appearance of birds.

The origin of feathers is a specific instance of the much more general question of the origin of evolutionary novelties—structures that have no clear antecedents in ancestral animals and no clear related structures (homologues) in contemporary relatives. Although evolutionary theory provides a robust explanation for the appearance of minor variations in the size and shape of creatures and their component parts, it does not yet give as much guidance for understanding the emergence of entirely new structures, including digits, limbs, eyes and feathers.

Progress in solving the particularly puzzling origin of feathers has also been hampered by what now appear to be false leads, such as the assumption that the primitive feather evolved by elongation and division of the reptilian scale, and speculations that feathers evolved for a specific function, such as flight. A lack of primitive fossil feathers hindered progress as well. For many years the earliest bird fossil has been *Archaeopteryx lithographica*, which lived in the late Jurassic period (about 148 million years ago). But *Archaeopteryx* offers no new insights on how feathers evolved, because its own feathers are nearly indistinguishable from those of today's birds.

Very recent contributions from several fields have put these traditional problems to rest. First, biologists have begun to find new evidence for the idea that developmental processes—the complex mechanisms by which an individual organism grows to its full size and form—can provide a window into the evolution of a spe-

OVERVIEW/FEATHER EVOLUTION

- The way a single feather develops on an individual bird can provide a window into how feathers evolved over the inaccessible stretches of prehistoric time. The use of development to elucidate evolution has spawned a new field: evolutionary developmental biology, or "evo-devo" for short.
- According to the developmental theory of feather origin, feathers evolved in a series of stages. Each stage built on an evolutionary novelty in how feathers grow that then served as the basis for the next innovation.
- Support for the theory comes from diverse areas of biology and paleontology. Perhaps the most exciting evidence comes from recent spectacular fossil finds of feathered dinosaurs that exhibit feathers at the various stages predicted by the theory.

cies' anatomy. This idea has been reborn as the field of evolutionary developmental biology, or "evo-devo." It has given us a powerful tool for probing the origin of feathers. Second, paleontologists have unearthed a trove of feathered dinosaurs in China. These animals have a diversity of primitive feathers that are not as highly evolved as those of today's birds or even *Archaeopteryx*. They give us critical clues about the structure, function and evolution of modern birds' intricate appendages.

Together these advances have produced a highly detailed and revolutionary picture: feathers originated and diversified in carnivorous, bipedal theropod dinosaurs before the origin of birds or the origin of flight.

THE TOTALLY TUBULAR FEATHER

This surprising picture was pieced together thanks in large measure to a new appreciation of exactly what a feather is and how it develops in modern birds. Like hair, nails and scales, feathers are integumentary appendages—skin organs that form by controlled proliferation of cells in the epidermis, or outer skin layer, that produce the keratin proteins. A typical feather features a main shaft, called the rachis. Fused to the rachis are a series of branches, or barbs. In a fractal-like reflection of the branched rachis and barbs, the barbs themselves are also branched: a series of paired filaments called barbules are fused to the main shaft of the barb, the ramus. At the base of the feather, the rachis expands to form the hollow tubular calamus, or quill, which inserts into a follicle in the skin. A bird's feathers are replaced periodically during its life through molt—the growth of new feathers from the same follicles.

Variations in the shape and microscopic structure of the barbs, barbules and rachis create an astounding range of feathers. But despite this diversity, most feathers fall into two structural classes. A typical pennaceous feather has a prominent rachis and barbs that create a planar vane. The barbs in the vane are locked together by pairs of specialized barbules. The barbules that extend toward the tip of the feather have a series of tiny hooklets that interlock with grooves in the neighboring barbules. Pennaceous feathers cover the bodies of birds, and their tightly closed vanes create the aerodynamic surfaces of the wings and tail. In dramatic contrast to pennaceous feathers, a plumulaceous, or downy, feather has only a rudimentary rachis and a jumbled tuft of barbs with long barbules. The long, tangled barbules give these feathers their marvelous properties of lightweight thermal insulation and comfortable loft. Feathers can have a pennaceous vane and a plumulaceous base.

In essence, all feathers are variations on a tube produced by proliferating epidermis with the nourishing dermal pulp in the center. And even though a feather is branched like a tree, it grows from its base like a hair. How do feathers accomplish this?

Feather growth begins with a thickening of the epidermis called the placode, which elongates into a tube—the feather germ. Proliferation of cells in a ring around the feather germ creates a cylindrical depression, the follicle, at its base. The growth of keratin cells, or keratinocytes, in the epidermis of the follicle—the follicle "collar"—forces older cells up and out, eventually creating the entire feather in an elaborate choreography that is one of the wonders of nature.

As part of that choreography, the follicle collar divides into a series of longitudinal ridges—barb ridges—that create the separate barbs. In a pennaceous feather, the barbs grow helically around the tubular feather germ and fuse on one side to form the rachis. Simultaneously, new barb ridges form on the other side of the tube. In a plumulaceous feather, barb ridges grow straight without any helical movement. In both types of feather, the barbules that extend from the barb ramus grow from a single layer of cells, called the barbule plate, on the periphery of the barb ridge.

EVO-DEVO COMES TO THE FEATHER

Together with various colleagues, we think the process of feather development can be mined to reveal the probable nature of the primitive structures that were the evolutionary precursors of feathers. Our developmental theory proposes that feathers evolved through a series of transitional stages, each marked by a developmental evolutionary novelty, a new mechanism of growth. Advances at one stage provided the basis for the next innovation.

In 1999 we proposed the following evolutionary scheme. Stage 1 was the tubular elongation of the placode from a feather germ and follicle. This yielded the first feather—an unbranched, hollow cylinder. Then, in stage 2, the follicle collar, a ring of epidermal tissue, differentiated (specialized): the inner layer became the longitudinal barb ridges, and the outer layer became a protective sheath. This stage produced a tuft of barbs fused to the hollow cylinder, or calamus.

The model has two alternatives for the next stage—either the origin of helical growth of barb ridges and formation of the rachis (stage 3a) or the origin of the barbules (3b). The ambiguity about which came first arises because feather development does not indicate clearly which event occurred before the other. A stage 3a follicle would produce a feather with a rachis and a series of simple barbs. A stage 3b follicle would generate a tuft of barbs with branched barbules. Regardless of which stage came first, the evolution of both these features, stage 3a+b, would yield the first double-branched feathers, exhibiting a rachis, barbs and barbules. Because barbules were still undifferentiated at this stage, a feather would be open pennaceous—that is, its vane would not form a tight, coherent surface in which the barbules are locked together.

In stage 4 the capacity to grow differentiated barbules evolved. This advance enabled a stage 4 follicle to produce hooklets at the ends of barbules that could attach to the grooved barbules of the adjacent barbs and create a pennaceous feather with a closed vane. Only after stage 4 could additional feather variations evolve, including the many specializations we see at stage 5, such as the asymmetrical vane of a flight feather.

THE SUPPORTING CAST

Inspiration for the theory came from the hierarchical nature of feather development itself. The model hypothesizes, for example, that a simple tubular feather preceded the evolution of barbs because barbs are created by the differentiation of the tube into barb ridges. Likewise, a plumulaceous tuft of barbs evolved before the pennaceous feather with a rachis because the rachis is formed by the fusion of barb ridges. Similar logic underlies each of the hypothesized stages of the developmental model.

Support for the theory comes in part from the diversity of feathers among modern birds, which sport feathers representing every stage of the model. Obviously, these feathers are recent, evolutionarily derived simplifications that merely revert back to the stages that arise during evolution, because complex feather diversity (through stage 5) must have evolved before *Archaeopteryx*. These modern feathers demonstrate that all the hypothesized stages are within the developmental capacity of feather follicles. Thus, the developmental theory of feather evolution does not require any purely theoretical structures to explain the origin of all feather diversity.

Support also comes from exciting new molecular findings that have confirmed the first three stages of the evo-devo model. Recent technological advances allow us to peer inside cells and identify whether specific genes are expressed (turned on so that they can give rise to the products they encode). Several laboratories have combined these methods with experimental techniques that investigate the functions of the proteins made when their genes are expressed during feather development. Matthew Harris and John F. Fallon of the University of Wisconsin-Madison and one of us (Prum) have studied two important pattern formation genes—*sonic hedgehog (Shh)* and *bone morphogenetic protein 2 (Bmp2)*. These genes play a crucial role in the growth of vertebrate limbs, digits, and integumentary appendages such as hair, teeth and nails. We found that Shh and Bmp2 proteins work as a modular pair of signaling molecules that, like a general-purpose electronic component, is reused repeatedly throughout feather development. The Shh protein induces cell proliferation, and the Bmp2 protein regulates the extent of proliferation and fosters cell differentiation.

The expression of Shh and Bmp2 begins in the feather placode, where the pair of proteins is produced in a polarized anterior-posterior pattern. Next, Shh and Bmp2 are both expressed at the tip of the tubular feather germ during its initial elongation and, following that, in the epithelium that separates the forming barb ridges, establishing a pattern for the growth of the ridges. Then in pennaceous feathers, the Shh and Bmp2 signaling lays down a pattern for helical growth of barb ridges and rachis formation, whereas in plumulaceous feathers the Shh and Bmp2 signals create a simpler pattern of barb growth. Each stage in the development of a feather has a distinct pattern of Shh and Bmp2 signaling. Again and again the two proteins perform critical tasks as the feather unfolds to its final form.

These molecular data confirm that feather development is composed of a series of hierarchical stages in which subsequent events are mechanistically dependent on earlier stages. For example, the evolution of longitudinal stripes in Shh-Bmp2 expression is contingent on the prior development of an elongate tubular feather germ. Likewise, the variations in Shh-Bmp2 patterning during pennaceous feather growth are contingent on the prior establishment of the longitudinal stripes. Thus, the molecular data are beautifully consistent with the scenario that feathers evolved from an elongate hollow tube (stage 1), to a downy tuft of barbs (stage 2), to a pennaceous structure (stage 3a).

THE STARS OF THE DRAMA

New conceptual theories have spurred our thinking, and state-of-the-art laboratory techniques have enabled us to eavesdrop on the cell as it gives life and shape to a feather. But plain old-fashioned detective work in fossil-rich quarries in northern China has turned up the most spectacular evidence for the developmental theory. Chinese, American and Canadian paleontologists working in Liaoning Province have unearthed a startling trove of fossils in the early Cretaceous Yi-xian formation (124 to 128 million years old). Excellent conditions in the formation have preserved an array of ancient organisms, including the earliest placental mammal, the earliest flowering plant, an explosion of ancient birdsd, and a diversity of theropod dinosaur fossils with sharp integumentary details. Various dinosaur fossils clearly show fully modern feathers and a variety of primitive feather structures. The conclusions are inescapable: feathers originated and evolved their essentially modern structure in a lineage of terrestrial, bipedal, carnivorous dinosaurs before the appearance of birds or flight.

The first feathered dinosaur found there, in 1997, was a chicken-size coelurosaur (*Sinosauropteryx*); it had small tubular and perhaps branched structures emerging from its skin. Next the paleontologists discovered a turkey-size oviraptoran dinosaur (*Caudipteryx*) that had beautifully preserved modern-looking pennaceous feathers on the tip of its tail and forelimbs. Some skeptics have claimed that *Caudipteryx* was merely an early flightless bird, but many phylogenetic analyses place it among the oviraptoran theropod dinosaurs. Subsequent discoveries at Liaoning have revealed pennaceous feathers on specimens of dromaeosaurs, the theropods, which are hypothesized to be most closely related to birds but which clearly are not birds. In all, investigators found fossil feathers from more than a dozen nonavian theropod dinosaurs, among them the ostrich-size therizinosaur *Beipiaosaurus* and a variety of dromaeosaurs, including *Microraptor* and *Sinornithosaurus*.

The heterogeneity of the feathers found on these dinosaurs is striking and provides strong direct support for the developmental theory. The most primitive feathers known—those of *Sinosauropteryx*—are the simplest tubular structures and are remarkably like the predicted stage 1 of the developmental model. *Sinosauropteryx*, *Sinornithosaurus* and some other nonavian theropod specimens show open tufted structures that lack a rachis and are strikingly congruent with stage 2 of the model. There are also pennaceous feathers that obviously had differentiated barbules and coherent planar vanes, as in stage 4 of the model.

These fossils open a new chapter in the history of vertebrate skin. We now know that feathers first appeared in a group of theropod dinosaurs and diversified into essentially modern structural variety within other lineages of theropods before the origin of birds. Among the numerous feather-bearing dinosaurs, birds represent one particular group that evolved the ability to fly using the feathers of its specialized forelimbs and tail. *Caudipteryx*, *Protopteryx* and dromaeosaurs display a prominent "fan" of feathers at the tip of the tail, indicating that even some aspects of the plumage of modern birds evolved in theropods.

The consequence of these amazing fossil finds has been a simultaneous redefinition of what it means to be a bird and a reconsideration of the biology and life history of the theropod dinosaurs. Birds—the group that includes all species descended from the most recent common ancestor of *Archaeopteryx* and modern birds—used to be recognized as the flying, feathered vertebrates. Now we must acknowledge that birds are a group of the feathered theropod dinosaurs that evolved the capacity of powered flight. New fossil discoveries have continued to close the gap between birds and dinosaurs and ultimately make it more difficult even to define birds. Conversely, many of the most charismatic and culturally iconic dinosaurs, such as *Tyrannosaurus* and *Velociraptor*, are very likely to have had feathered skin but were not birds.

A FRESH LOOK

Thanks to the dividends provided by the recent findings, researchers can now reassess the various earlier hypotheses about the origin of feathers. The new evidence from developmental biology is particularly damaging to the classical theory that feathers evolved from elongate scales. According to this scenario, scales became feathers by first elongating, then growing fringed edges, and finally producing hooked and grooved barbules. As we have seen, however, feathers are tubes; the two planar sides of the vane—in other words, the front and the back—are created by the inside and outside of the tube only after the feather unfolds from its cylindrical sheath. In contrast, the two planar sides of a scale develop from the top and bottom of the initial epidermal outgrowth that forms the scale.

The fresh evidence also puts to rest the popular and enduring theory that feathers evolved primarily or originally for flight. Only highly evolved feather shapes—namely, the asymmetrical feather with a closed vane, which did not occur until stage 5—could have been used for flight. Proposing that feathers evolved for flight now appears to be like hypothesizing that fingers evolved to play the piano. Rather feathers were "exapted" for their aerodynamic function only after the evolution of substantial developmental and structural complexity. That is, they evolved for some other purpose and were then exploited for a different use.

Numerous other proposed early functions of feathers remain plausible, including insulation, water repellency, courtship, camouflage and defense. Even with the wealth of new paleontological data, though, it seems unlikely that we will ever gain sufficient insight into the biology and natural history of the specific lineage in which feathers evolved to distinguish among these hypotheses. Instead our theory underscores that feathers evolved by a series of developmental innovations, each of which may have evolved for a different original function. We do know, however, that feathers emerged only after a tubular feather germ and follicle formed in the skin of some species. Hence, the first feather evolved because the first tubular appendage that grew out of the skin provided some kind of survival advantage.

Creationists and other evolutionary skeptics have long pointed to feathers as a favorite example of the insufficiency of evolutionary theory. There were no transitional forms between scales and feathers, they have argued. Further, they asked why natural selection for flight would first divide an elongate scale and then evolve an elaborate new mechanism to weave it back together. Now, in an ironic about-face, feathers offer a sterling example of how we can best study the origin of an evolutionary novelty: focus on understanding those features that are truly new and examine how they form during development in modern organisms. This new paradigm in evolutionary biology is certain to penetrate many more mysteries. Let our minds take wing.

DINOSAUR OR BIRD? THE GAP NARROWS

As this issue of *Scientific American* went to press, researchers announced a startling new find in China: a dinosaur with asymmetrical feathers, the only kind of feathers useful for flight, on its arms and legs. Before this discovery, scientists had thought that birds were the only creatures that possessed asymmetrical feathers. In fact, such feathers were one of the few unique characteristics that distinguished the avian descendants from their dinosaur forebears. Now it appears that even flight feathers, not merely feathers per se, existed before birds.

Writing in the January 23 issue of *Nature*, Xing Xu, Zhonghe Zhou and their colleagues from the Institute of Vertebrate Paleontology and Paleoanthropology of the Chinese Academy of Sciences report that a newly discovered species of *Microraptor* had modern-looking asymmetrical flight feathers creating front and hind "wings." Moreover, the feathers are more asymmetrical toward the end of the limb, just as occurs on the modern bird wing.

Debate on the origin of bird flight has focused on two competing hypotheses: flight evolved from the trees through an intermediate gliding stage or flight evolved from the ground through a powered running stage. Both have good supporting evidence, but Xu and his colleagues say the new *Microraptor* find furnishes additional support for the arboreal hypothesis because together the forelimb and leg feathers could have served as a "perfect airfoil." Substantial questions remain of course, among them, How did *Microraptor* actually use its four "wings"?

—The Editors

RICHARD O. PRUM *and* **ALAN H. BRUSH** *share a passion for feather biology. Prum, who started bird-watching at the age of 10, is now associate professor of ecology and evolutionary biology at the University of Kansas and curator of ornithology at the Natural History Museum and Biodiversity Research Center there. His research has focused on avian phylogeny, avian courtship and breeding systems, the physics of structural colors, and the evolution of feathers. He has conducted field studies in Central and South America, Madagascar and New Guinea. Brush is emeritus professor of ecology and evolutionary biology at the University of Connecticut. He has worked on feather pigment and keratin biochemistry and the evolution of feather novelties. He was editor of* The Auk.

MORE TO EXPLORE

Development and Evolutionary Origin of Feathers. Richard O. Prum in *Journal of Experimental Zoology (Molecular and Developmental Evolution)*, Vol. 285, No. 4, pages 291–306; December 15, 1999.

Evolving a Protofeather and Feather Diversity. Alan H. Brush in *American Zoologist*, Vol. 40, No. 4, pages 631–639; 2000.

Rapid Communication: Shh-Bmp2 Signaling Module and the Evolutionary Origin and Diversification of Feathers. Matthew P. Harris, John F. Fallon and Richard O. Prum in *Journal of Experimental Zoology*, Vol. 294, No. 2, pages 160–176; August 15, 2002.

The Evolutionary Origin and Diversification of Feathers. Richard O. Prum and Alan H. Brush in *Quarterly Review of Biology*, Vol. 77, No. 3, pages 261–295; September 2002.

Bird's-Eye View[*]

By Matthew T. Carrano and Patrick M. O'Connor
Natural History, May 2005

The array of the dinosaurs that flourished during the Mesozoic era was as dazzling as any bestiary ever imagined; not even medieval fantasies of griffins and unicorns could compete with the fabulous record of fossils in rock. Yet with a single exception, the entire dinosaur lineage was obliterated 65 million years ago. The sole dinosaurian representatives to survive the cataclysm were the birds, a group that has since radiated into virtually every environment on the planet.

The suggestion of an evolutionary link between dinosaurs and birds originated with several late-nineteenth-century biologists, most notably Darwin's friend Thomas Henry Huxley. At first welcomed, the hypothesis was later disregarded by most biologists and treated with skepticism through much of the twentieth century. But in the past three decades, the hypothesis has roared back to life, with almost overwhelming support. The latest evidence for the link has come from the spectacular recent discoveries of a number of feathered dinosaurs in China.

To many a casual eye, the case is made by the presence of feathers on the fossils. But feathers only highlight one of the most visible similarities between the two groups. Biologists classify birds among the dinosaurs not only because both groups have (or had) feathers, but also because they share a suite of other, characteristic anatomical traits. One of those important traits is the "pneumaticity" of the skeleton: certain dinosaurs possessed bones riddled with air pockets, which during life were linked with the pulmonary, or breathing, system of the animal. Much the same is the case with many birds today.

The classification of birds as dinosaurs also implies that many other so-called avian features are better thought of as dinosaurian. And similar anatomies could imply that the bodies of birds and dinosaurs functioned similarly. Moreover, one may also learn a great deal about dinosaur biology by contrasting their features with the anatomical and biomechanical characteristics of other, more distantly related vertebrates. It is the birds, though, that have carried the torch of dinosaurian biological heritage from the Mesozoic through global calamity to the present day. Modern paleontologists, in large part by the light of that torch, are elucidating the paleobiological characteristics of those long-dead, long-buried, long-obscured animals.

To understand what one can learn about dinosaurs from the study of birds, it is useful to sketch how the two groups are related. A discipline of biology known as cladistics, or phylogenetic systematics, investigates the evolutionary relationships among organisms by charting their anatomical similarities. Cladistic hypotheses about such interrelations often take the form of a branching diagram called a cla-dogram. Each junction on the cladogram indicates an evolutionary event that split one lineage into two. Each of the two descendant lineages shares one or more features inherited from the ancestor at the most recent junction, and those shared features define different groups. To examine the relations within and between groups of organisms is also to chronicle the sequence by which those groups' features evolved.

According to the leading cladistic hypotheses, birds are descended from within the group of theropod dinosaurs. Theropods are quite familiar to most people, if not necessarily by that name: members include giant *Tyrannosaurus*, sickle-clawed *Velociraptor*, and birdlike *Ornithomimus*. Theropods such as *Herrerasaurus*, from the Middle Triassic, are among the earliest known dinosaurs.

Theropods, like birds, were bipedal animals. All of them share several key features: thin-walled bones, a foot with three main toes, and a joint in the lower jaw. Early theropods split into two groups, the herrerasaur-like primitive theropods, and a group called the neotheropods, which included most of the familiar predatory dinosaurs. Early neotheropods, known as the coelophysoids, were common in the Late Triassic and Early Jurassic.

As the neotheropods emerged as a separate group, they shared an important "birdlike" trait—the furcula, often (in birds) called the wishbone. The furcula is formed by the fused left and right clavicles, and in modern birds it acts as a spring between the powerful flapping wings. Clearly, though, the furcula did not function in that capacity in the earliest neotheropods. Although its original role remains unclear, it may have helped neotheropods control their forelimbs.

By the end of the Early Jurassic the theropods split again, giving rise to the cera-tosaurs (a group that includes *Ceratosaurus*) and tetanurans (a diverse group that includes *Allosaurus*, *Spinosaurus*, *Tyrannosaurus*, *Velociraptor*, and a number of others). The tetanurans are named for their tails, which were less flexible than those of their forebears. Like the hand of a modern bird, the tetanuran hand had only three fingers; the tetanurans' wrist was more specialized, and their entire forelimb more birdlike, than the corresponding anatomy of any of the earlier theropods.

Around the same time the allosaurs appeared, another subgroup of the teta-nurans, the coelurosaurs, also branched off. Coelurosaurs included both large species, such as *Tyrannosaurus rex*, and small ones, some not much bigger than a chicken. The coelurosaurs—particularly their subgroup known as maniraptorans (to which *Velociraptor* and many other dinosaur species belonged)—show the greatest affinities with birds. Some early forms, including primitive tyrannosaurs, had a downy covering on the skin, possibly either for insulation or for display. Other species had distinct feathers covering nearly the entire body. Maniraptorans also had a specialized shoulder blade and a unique, curved bone in the wrist, which enabled the hand to move in just one plane. The motion was similar to wing folding in modern birds.

Finally, with just a few additional modifications—such as the lengthening of the forearm and hand—the last living subgroup of the maniraptorans appeared: the true birds.

The hypothesized interrelations expressed by a cladogram can guide paleontologists to specific evolutionary patterns that can shed light on other aspects of dinosaurian biology. For example, how did dinosaurs move? Living animals, of course, confront the same laws of physics as the dinosaurs did. By studying the biomechanics of locomotion in living animals, then, dinosaur biologists can focus more precisely on what the fossil evidence can convey.

Early theropod locomotion was not particularly specialized—theropods were, in general, neither runners nor plodders, neither climbers nor diggers nor swimmers; more likely, they were jacks of many of those trades, but masters of none. Their most notable attribute was an inherited one: bipedalism. The original dinosaurs walked on two legs, making the group an oddity in the history of vertebrates. In spite of their shared bipedalism, various theropod groups did become more specialized in their locomotion, as their skeletons attest. Comparing their bones with those of modern animals can help show how differences in anatomy translate into differences in behavior.

Among living groups, the distal—that is, distant from the body's center—segments of limbs are relatively long (compared with the rest of the body) in fast runners such as ostriches and cheetahs, and in long-distance runners such as wildebeest and caribou. Animals with relatively short distal bones, such as elephants and hippopotamuses, have more columnar legs and do little running. Between those extremes is a near-continuum of variation. The proportions of the distal limb bones in theropods were generally intermediate between the extremes of cheetah and elephant.

Another mammalian tendency is that large species typically have relatively short limbs and small species relatively long ones. The same pattern held in theropods. Some large theropods, such as spinosaurs and allosaurs, had lionlike limbs—perhaps because they hunted by stealth or covered relatively little ground in their roamings. Other species bucked the trend, though. Tyrannosaurs had a runner's limbs, despite their enormous size—indicating that they were probably adapted to running relatively fast or far.

Another way to examine theropod biomechanics is to reconstruct the musculature of the limbs. Muscle-attachment marks on fossilized theropod bones can be compared to similar marks on the dinosaur's nearest living relatives: the crocodilians—whose ancestors were quadrupedal—and the birds.

The hind-limb muscles of birds are well adapted for bipedal motion. The muscle arrangement at the hip and knee maximizes stability, yet gives the leg the ability to make wide swings fore and aft. But bipedalism in birds is a highly specialized form of bipedal motion; the large tails of birds' ancestors, which in crocodilians still anchor the leg muscles, have mostly vanished in birds. Hence, in birds, the muscles attached to the tail are also small. Birds walk in a crouched posture, moving the knee more than the hip—what biologists sometimes call "Groucho running," after Groucho Marx.

An analysis of theropod fossils shows that the animal had birdlike limb muscles early in their evolution. That was certainly the case by the time the coelophysoids appeared, and perhaps even by the time of *Herrerasaurus*. The repositioning of the muscles changed the way theropods walked: they began moving their legs as birds do—in one plane, fore and aft—rather than as crocodiles do, waddling from side to side.

In later theropods, such as allosaurs and tyrannosaurs, several new muscle attachments appeared, which occur in birds but not in crocodilians or earlier theropods. Yet despite the rearrangement of the attachments of some leg muscles, most theropods still retained substantial attachments of the leg muscles to the tail. The Mesozoic world was probably not full of Groucho-running theropods. Rather, the leg muscles attached to the tail would have caused the upper part of the limb to move at the hip.

Although great size, as well as a great range of body sizes, are among the most familiar qualities of dinosaurs, the early theropods were both small and fairly uniform. *Eoraptor*, one of the earliest theropods, was perhaps three feet long and weighed about twenty-five pounds, more or less the dimensions of a medium-size dog. Yet even that animal was much larger than its nearest ancestors. Further change came quickly. By the Late Triassic, the dominant predators were coelophysoid theropods, a group ranging from the nine-foot-long *Syntarsus* to the fifteen-foot-long *Gojirasaurus*. But the first large theropods, animals more than thirty feet long and weighing between two and three tons, appeared during the Late Jurassic. The true giants did not arrive until the middle of the Cretaceous period. The carcharodontosaurs were among the largest terrestrial predators that ever lived, some reaching as much as forty feet long and weighing four tons. The spinosaurs were contemporary to the carcharodontosaurs, and similar in size. The giant tyrannosaurs appeared in the latest part of the Cretaceous, reaching or exceeding the carcharodontosaurs in size.

What is interesting is not so much the absolute size of those giant predators but that at least three lineages of theropods independently evolved to almost exactly the same size. Something, it would seem, made such a size advantageous. Or perhaps something structural or ecological made any larger size a real disadvantage.

Body size affects nearly every aspect of organismal biology. The basic physics of size dictates an animal's structure and function in a number of predictable ways. For example, when an animal doubles its linear dimensions, its volume increases eightfold. Hence processes that depend on volume, such as maintaining body temperature, are highly sensitive to changes in body size. Other physiological processes that depend on surface areas—gas exchange across a membrane, for instance—are intermediately affected by changes in body size. One consequence of those geometric relationships is that the larger the animal, the harder it becomes to adjust its body temperature. Body temperature is regulated through the body's surface area, but heat is stored in the body's volume.

To some degree, most living reptiles rely on the external environment for controlling body temperature; thus, reptiles are called ectothermic, or, somewhat erroneously, "cold-blooded." Birds, however, can fine-tune their body temperatures internally, a condition referred to as endothermy, or, also somewhat erroneously, "warm bloodedness." But the apparent dichotomy between endothermy and ectothermy is misleading; rather, there is a broad spectrum of metabolic types, many of which are directly correlated with the anatomical form and function of the breathing apparatus.

Many reptiles have relatively simple lungs. As they expand or contract, air flows in and out of them through the same channels, just as it does in people. But the configuration of the internal cavity of the reptilian lung varies from species to species. In a few species, including sonic lizards, each lung is a simple sac, and gases are exchanged only around its edges. In other species, such as monitor lizards and crocodilians, the lungs are partitioned into chambers made up of an intricate net of support structures. The network provides a larger surface area, which enables higher rates of gas exchange than do the edges of a simple, saclike lung.

Modern birds have modified the basic reptilian design in such a way as to increase lung partitioning, and, consequently, the surface area for gas exchange. But unlike the reptilian lung, the avian lung changes very little in size during ventilation. Instead, birds have flexible air sacs (usually nine in number) that act as bellows to move air through the lungs. Although the air sacs are connected to the lungs, they do not take part directly in gas exchange. Furthermore, in some parts of the bird lung, air flows almost continually in just one direction. Those specializations enable birds to exchange gases efficiently enough to sustain high metabolic rates and regulate their temperatures internally.

Although avian lungs and air sacs are made of soft tissues, they have important connections with the skeleton. Extensions from the air sacs physically invade the skeleton, a process known as pneumatization. The result can be dramatic. Imagine walking down the windpipe of a bird, into its lungs, and then on into the skeleton, including the backbone and limbs—within the pulmonary system. That's some fantastic voyage!

No other living animals have pneumatic bones like those of birds, but substantial evidence suggests that theropods, along with the flying pterosaurs and sauropod dinosaurs, had at least a superficially similar pulmonary system. Like birds, those animals had holes in the outer surfaces of many of their bones. The holes were connected to large, air-tilled chambers within the bones.

Even in birds, though, the function of pneumatic bones remains unclear. No gases are exchanged within the bones, nor do the air-filled chambers in the bones help move air through the lung—bone, after all, is not a flexible bellows. One plausible idea is that pneumatic bones might have evolved because they replaced heavy (and metabolically expensive) bone marrow with air. Pneumatic bones enable a bird (or a dinosaur) to expand its overall body size without a commensurate increase in weight.

In spite of the uncertain role of pneumatic bones, their presence in theropods suggests that at least some theropods had air sacs similar to those observed in birds. Without additional evidence, though, it is probably idle to speculate any further about how theropods breathed. Nevertheless, the historical perspective provided by theropod pneumaticity may be the key to understanding the origin of air-filled, lightweight bones in birds.

Ornithologists have long sought to explain pneumatic bones in birds as an adaptation to some aspect of their lifestyle, such as the great benefit they offer for energy savings in flying. Pneumaticity clearly originated much earlier in avian history, but perhaps for a similar reason—that is, its adaptive value in relaxing the constraints on the size of theropods. One point corroborating that idea is that many of the largest theropods, such as carcharodontosaurs and tyrannosaurs, often had the most extreme pneumaticity. Many of the smaller theropods, in contrast, only pneumatized certain regions of the vertebral column.

Similar patterns of pneumaticity occur in birds: among flying birds, at least, the larger the bird, the more extensive its pneumaticity. Certain large-bodied flying birds, such as bustards, pelicans, and vultures, pneumatize virtually the entire skeleton, out to the tips of the wings. Many medium-size and small birds, such as ducks, pheasants, and songbirds, only pneumatize the vertebrae and limb bones closest to the lung and air sacs. Some interesting exceptions to the correlation between body size and pneumaticity occur in birds that dive underwater to feed, such as grebes, loons, and penguins. Those birds have eliminated bony pneumaticity altogether, so as to reduce their buoyancy when they dive.

The broad variation in skeletal pneumaticity among birds suggests that interactions between the pulmonary and skeletal systems alter drastically in response to a variety of physical and environmental pressures. Could similar variations in pneumaticity reflect the various physical and ecological factors theropods had to confront? With birds as a model, paleontologists should be able to frame and test hypotheses that can begin to answer that question.

Using living animals as "model organisms" for understanding dinosaur biology offers many advantages over traditional methods of paleontology. But paleobiologists must also remain cautious when making inferences related to the activities of long-dead animals. For example, as tempting as it is to read "bird" into every dinosaurian trait, it is just as important to acknowledge the limits of current knowledge, and the fact that the dinosaurs maintained their own evolutionary trajectory; they likely possessed an amalgam of traits present in modern birds and their reptilian relatives.

Ideally, paleontology integrates multiple lines of evidence, from a variety of living and extinct animals, to assess the full biological potential of long-extinct groups. That approach is not without its limits. Nevertheless, by seeking novel ways to integrate the vast array of biological subdisciplines, paleobiologists, are beginning to put a modern face on some very old "terrible lizards." Those complementary studies will ultimately provide the most rigorous assessment of how dinosaurs actually lived.

Dinotopia[*]

By Jeff Hecht
New Scientist, May 21, 2005

Mei long, the "soundly sleeping dragon", is small enough to hold in your two hands. Its head is tucked under its forelimb, like a sleeping bird with its head under a wing. When I saw it in Xu Xing's office at the American Museum of Natural History in New York, the long tail that was wrapped around its body when it was discovered had been removed for analysis. Otherwise the tiny bones of this little dinosaur were arranged just as they were when it went to sleep for the last time, some 125 million years ago in what is now China.

Such a complete, exquisitely preserved skeleton of a small dinosaur was something that palaeontologists could only have dreamed about a decade ago. In recent years, however, they have become rather accustomed to dream discoveries. *Mei long* is just the latest in a string of stunning finds to come out of a rock formation called Yixian in the Liaoning province of north-eastern China. The Yixian formation is stuffed full of dinosaurs, as well as birds, mammals, lizards, fish, turtles, insects and plants, all preserved in great detail. And over the past 10 years, this fossil bonanza has helped to transform our knowledge and understanding of dinosaurs, the world they inhabited, and the animals they shared it with.

Well before *Mei long* turned up, Yixian was well known for its beautifully preserved fossilised fish, insects and turtles, which were—and still are—collected and sold by peasant farmers. The dinosaur story, however, only began in earnest about a decade ago, when word began circulating that Chinese scientists had found an amazing little predator named sinosauropteryx.

Sinosauropteryx—meaning "Chinese winged lizard"—made its informal world debut in October 1996 at the Society of Vertebrate Paleontology's annual meeting at the American Museum of Natural History. It wasn't on the programme, but Phil Currie of the Royal Tyrell Museum in Drumheller, Canada, brought snapshots from China. Circles of scientists gazed in amazement at the photos: the creature was clearly covered in what looked like downy feathers. Among the onlookers was Yale University's John Ostrom, the grand old man of dinosaur palaeontology. Three decades earlier, Ostrom had proposed that birds evolved from dinosaurs. By 1996 his walk was slow and his hair was white, but when he saw the pictures of sinosauropteryx his eyes lit up.

It was a marvellous specimen. Its body had fallen to the bottom of a lake, probably during a volcanic eruption, where sediments flattened it, preserving a perfect impression of the little dinosaur—including a fringe of feathery filaments around its body. Ostrom had not believed he would live to see a feathered dinosaur. Neither had the much younger Currie. Both were wrong.

More soon followed. In early 1997 The Academy of Natural Sciences in Philadelphia, Pennsylvania, arranged a whirlwind tour of China for Ostrom and three others: Larry Martin of the University of Kansas, pterosaur expert Peter Wellnhofer of Ludwig Maximilian University of Munich, Germany, and feather-development specialist Alan Brush of the University of Connecticut. After showing his guests the sinosauropteryx specimen, Ji Qiang of the National Geological Museum in Beijing stunned them with two additional feathered dinosaurs that no one in the west knew existed. He called one protarchaeopteryx because it looked like a primitive version of the oldest-known bird, archaeopteryx. Another he called caudipteryx, meaning "winged tail".

Up to that point, feathers were thought to be unique to birds. Dinosaurs were believed to have been scaly: the few impressions of dinosaur skin that had been found all showed reptilian scales. And while microscopic examination showed that sinosauropteryx's frill was made of filaments rather than real feathers, both protarchaeopteryx and caudipteryx had the real thing. Suddenly fully fledged feathers were no longer strictly for the birds; dinosaurs had feathers, too.

Brains buzzing, the four headed home. Ostrom declared the Yixian fossil beds held a century's worth of work. Soon palaeontologists from all over the world were packing their bags for China, hoping to find fossils that would solve old mysteries and present new ones. A few years on, they are ready to start saying what it all means.

Fossils are failures of nature's recycling system. Big dinosaur bones survive because of their sheer size, but the smaller bones are likely to be lost, and teeth are often the only traces of small animals. Soft parts such as skin and feathers almost never fossilise. Yet once in a very great while, nature slips up and leaves a treasure trove of fossils preserved in exquisite detail. The Burgess Shale in British Columbia, Canada, is arguably the most famous of these, preserving strange soft-bodied creatures from the Cambrian explosion, the first great blossoming of animals more than 500 million years ago. The limestone of Solnhofen in Bavaria, Germany, where archaeopteryx was discovered in 1860, is another.

Now the Yixian formation has joined the pantheon—and arguably has topped the lot. "It's the biggest fossil bonanza ever," says Martin. A handful of other sites preserve some exceptional fossils from the Mesozoic era, the age of the dinosaurs, but none has the diversity of the Yixian formation. It reveals "almost every group of animals we would expect to see", says Paul Barrett of the Natural History Museum in London. Only crocodiles are missing. No one knows why, since they are abundant in Solnhofen and other Mesozoic sites.

Spread across a broad area of north-eastern China, the Yixian formation is in an area of semi-arid hills criss-crossed by streams that cut through the sedimentary rocks and expose the deposits. In many respects it resembles the fossil-rich badlands of western North America. When sinosauropteryx and its kin roamed the region 125 million years ago, it was covered by forest and quiet freshwater lakes. Nearby volcanoes erupted from time to time, spreading thick layers of ash that killed many animals. Birds may have succumbed to fumes from the eruptions and fallen into the lake; suffocated dinosaurs may have sunk to the bottom when swimming or after being washed into the water. Some palaeontologists think that volcanism may have occasionally released lethal clouds of carbon dioxide above the lake, like those of Lake Nyos in Cameroon, downing flocks of birds. Today, the fossil beds consist of alternating layers of volcanic ash and lake-bed sediments that are well over a kilometre thick in places. So far they have yielded more than 20 new dinosaurs, and it is anyone's guess how many more remain to be discovered.

The spectacular fossils have kept scientists so busy that they have yet to figure out the details of the fossilisation process. But it is clear there are two different processes at work, both linked with eruptions—and both remarkable. Some dead animals and plants found their way into a deep lake and sank to the bottom where volcanic ash mixed with lake sediments to preserve flattened imprints like that of sinosauropteryx. Others were buried alive in fine ash, in a dinosaurs' Pompeii, producing 3D fossils such as *Mei long*. The two types are complementary. Ash fossils preserve the 3D shape but not the soft parts; lake-bed fossils flatten skeletons, but preserve details such as soft tissue and body coverings.

Initially, Chinese geologists believed the Yixian rocks formed in the late Jurassic period, which ended about 145 million years ago. But those dates were only estimates based on comparing the fossils with those found elsewhere. When geologists dated the rocks radiometrically—the gold standard for geologic age—they found that the oldest rocks were just 127 to 128 million years old, with some as young as 121 million years. That placed them in the early Cretaceous period, the final and most spectacular chapter in the age of the dinosaurs. The bulk of the fossils date from 127 to 125 million years ago, says Barrett.

The fossil beds contain an astounding diversity of life. They preserve the oldest known flowers, illuminating the birth of flowering plants. They also include mammals—some with fur—at a key time in the evolution of modern groups. Among these are stunning surprises like *Repenomamus giganticus*, a badger-sized mammal that dined on dinosaurs at a time when mammals were supposed to be tiny, timid and nocturnal. But amid all this diversity it is the little dinosaurs that have attracted the most attention—and not just because they are feathered.

One of their principal attractions is their size. Big dinosaurs might get the public's pulse racing, but as far as evolutionary biologists are concerned the little guys are where the action is. Small animals tend to be highly adaptable generalists that go on to evolve specialist traits and then expand in size.

Before Yixian, no one doubted that small dinosaurs had existed. Few had been found, though. There were teeth, some scattered bones, and a handful of partial skeletons. But it wasn't clear how many of the little fossils were youngsters of species that we already knew about. The Yixian fossils have changed all that, and the evolutionary biologists have not been disappointed. Xu says most of the evolutionary innovation at Yixian seems concentrated in the smaller dinosaurs. Many are the earliest known members of groups that went on to dominate the Cretaceous, from triceratops to *Tyrannosaurus rex*.

That's not to say that there were no big dinosaurs in Yixian 125 million years ago, but their remains are few and far between. Perhaps they didn't float far enough from shore to be dropped deep in the ancient lakes, and so never reached the fossil-forming sediments. Another possibility is that the local fossil-hunting fanners are not looking for big animals. Palaeontologists have found the remains of an elephant-sized plant-eater that appears to be related to diplodocus and other long-necked browsers, though it has yet to be properly described. "The farmers chucked it over the hillside," Currie says.

The single most common small dinosaur in the Yixian formation is psittacosaurus, a parrot-beaked plant-eater that walked on two legs. Psittacosaurus belongs to the large group of plant-eating dinosaurs called ornithischians, which included the later duck-billed dinosaurs. Adults grew to be a couple of metres long and had large numbers of young; one was found incubating a nest of over 30 juveniles. Thousands of psittacosauruses have been found in both sediments and ash, and the little ones were the favourite meals of local predators, including the badger-sized mammal repenomamus. But the psittacosauruses were doing something right; their descendants include the horned dinosaurs like triceratops and the dome-headed pachycephalosaurs.

Most of the more than 20 species of Yixian dinosaurs are theropods, a big class of bipedal dinosaurs typified by velociraptor. The Yixian theropods are strikingly diverse, ranging from the clearly dinosaurian, such as sinosauropteryx, to the very bird-like, such as caudipteryx. Most were predators but at least two—beipiaosaurus, a shaggy 3-metre beast that was the largest of Yixian's feathered dinosaurs, and the curiously buck-toothed incisivosaurus—were plant-eaters. Most ran on two legs, with two grasping arms in front and a long tail behind. They all had feathery coats. True birds are theropods too, and some 20 of them have been discovered at Yixian.

The ubiquity of feathers among the Yixian theropods was a huge surprise. Palaeontologists had long suspected that some theropods would bear feathers—after all, the consensus has long been that birds evolved from small, predatory theropods. But they never thought they would see feathers on almost everything.

One such surprise was a creature called dilong, a 1.5-metre theropod discovered by Xu and Mark Norell, who is also at the American Museum of Natural History. Dilong was covered in 2-centimetre-long filamentary feathers. It came as a surprise, then, when it turned out to be a forerunner of the tyrannosaurs. "I never expected tyrannosaurs would have feathers," says Norell. Adult *T. rexes* apparently did not have feathers; the handful of skin impressions discovered so far show scales. However, juveniles may have retained a downy coat of feathers to keep them warm.

Feathers are now so established on the dinosaur scene that when palaeontologists discovered the bones of a small theropod called bambiraptor in Cretaceous rocks in Montana, they simply assumed it was covered in feathers. Likewise *Mei long*: the ash that buried the sleeping dragon did not preserve its skin, but nobody doubts it was feathered.

As yet, only theropods are known to have had feathers. But there are hints that skin coverings might be more common than previously thought. Some fossils of the parrot-beaked psittacosaurus have frills of filaments along the tops of their tails.

And with all these feathers flying around, one obvious question is what the Yixian fossils tell us about the evolution of birds and the origin of flight.

In the 19th century Thomas Huxley pointed out that the skeleton of archaeopteryx resembled the only dinosaur known from Solnhofen, a small theropod named compsognathus. But at the time it was hard to envisage how supposedly sluggish, coldblooded dinosaurs could have evolved into active, warm-blooded birds. A century later Ostrom revived the question by showing striking similarities between archaeopteryx and deinonychus, one of the dromaeosaurs—a family of lightly built, agile carnivores that includes velociraptor. He suggested that birds had evolved from fast-running feathered dinosaurs that eventually became airborne.

More evidence that birds had evolved from dinosaurs emerged from a technique called cladistics, which builds evolutionary trees based on shared traits: the more traits two species share, the closer their relationship. This kind of analysis usually places birds within the theropods, close to two groups of dinosaurs: dromaeosaurs and troodonts, the group to which *Mei long* belongs.

By the early 1990s, most specialists were convinced and began calling birds "avian dinosaurs", which meant traditional dinosaurs had to be rebranded as "non-avian dinosaurs". However, a small but vocal group, including Martin, disagreed. They believed that birds evolved from tree-climbing non-dinosaur reptiles that evolved flight by gliding down to the ground. One of their best arguments was that it wasn't obvious how "ground-up" flight could have evolved, as runners would not benefit from the series of slight increases in arm and feather size needed to evolve wings. In contrast, even a small increment would have benefited early gliders as they jumped down from the trees.

The Yixian fossils injected new life into the debate. Here was a deposit that contained both birds and their dinosaur relatives. And, even better, they all had feathers.

The early Yixian discoveries appeared to confirm the idea that birds had evolved from fast-running theropods. Their feathers, for example, became more bird-like the closer they were to birds on the cladistic trees. Sinosauropteryx had simple filaments; protarchaeopteryx had feathers very similar to those on a modern bird. But then a tiny dinosaur fossil called microraptor threw a spanner in the works.

Microraptor's beginnings were inauspicious. Back in 1999 Stephen Czerkas, who runs The Dinosaur Museum in Blanding, Utah, was at the annual Tucson Gem and Mineral Show in Arizona. He spotted a beguiling Yixian fossil that seemed to be a vital missing link in bird evolution. Its front half was clearly bird-like, but the back half bore the unmistakable long stiff tail of a dromaeosaur—the fast-running ground dwellers from which birds supposedly evolved. Thinking he had an important discovery, he bought it, named it archaeoraptor, and tried to publish a scientific paper on it.

Archaeoraptor turned out to be a fake. As with the majority of Yixian fossils, it was found not by professional palaeontologists but by one of the local farmers who supplement their meagre income by selling fossils. And as with many fossils, it had been enhanced to increase its value. When Currie and Xu examined it they discovered that the fossil was exactly what it looked like: the front half of a bird with the back half of a dinosaur.

But it was still a great find. It turned out that the tail came from a previously unknown theropod of immense scientific value, which Xu and Zhonghe Zhou of the Institute of Vertebrate Palaeontology and Palaeoanthropology in Beijing named microraptor. It was the smallest known dinosaur: the whole animal was just 39 centimetres long, and 24 centimetres of that was tail. Xu and Zhou classed it as the most primitive known dromaeosaur, which made it hugely relevant to the bird evolution debate.

The first microraptor fossils didn't have feathers, but two years ago Xu and Zhou stunned palaeontologists with six specimens of a closely related species, Microraptorgui, which had long feathers on its arms, hind legs and tail. The little dinosaur's arms were clearly too weak and short for powered flight, and so Xu suggested it was a glider that coasted down from the trees by spreading its forewings while holding its hind legs and tail together to form an aerofoil. In other words, the orthodoxy was only half right. Birds were indeed descended from dinosaurs, but flight evolved from the trees down.

Of course, microraptor itself couldn't be the ancestor of birds—the Yixian formation is younger than archaeopteryx. But perhaps the two shared a common ancestor. The discovery of microraptor's leg feathers prompted palaeontologists to re-examine other fossils. Amazingly, archaeopteryx and some other early birds turned out to have leg feathers that palaeontologists hadn't noticed before. So did pedopenna, an incomplete bird-like theropod from the Daohugou fossil beds of the Inner Mongolia region of China. Those beds are not well dated, but are somewhat older than the Yixian formation. Both Xu and Zhou now argue that there was a glider stage on the road to powered flight.

Many other palaeontologists are sceptical or reserve judgement. "How you interpret it is open for discussion," says Currie. Tom Holtz of the University of Maryland is also cautious, as is Kevin Padian of the University of California, Berkeley. Martin, meanwhile, believes that microraptor is part of the story of flight, although he considers it and other dromaeosaurs to have been flightless birds rather than dinosaurs.

Origin of Species[*]

How a *T. Rex* Femur Sparked a Scientific Smackdown

By Evan Ratliff
Wired, July 2009

Sixty-eight million years ago, on a soggy marsh in what is now a desolate stretch of eastern Montana, a *Tyrannosaurus rex* died. In 2000, a team of paleontologists led by famed dinosaur hunter Jack Horner found it.

These are scientific facts, as solid as the chunk of fossilized femur from that same *T. rex* that Horner gave to North Carolina State University paleontologist Mary Schweitzer in 2003. It was labeled sample MOR 1125.

Several facts concerning MOR 1125 are also beyond dispute: First, that a technician in Schweitzer's lab put a piece of the bone in a demineralizing bath to study its components but left it in longer than necessary; when she returned, all that remained was a pliable, fibrous substance. That Schweitzer, intrigued by this result, ground up and prepared another piece of the bone and sent it to John Asara, a mass spectrometry expert at Beth Israel Deaconess Medical Center and Harvard Medical School. That Asara treated the brown powder with an enzyme and injected it into a mass spectrometer the size of a washing machine, hoping to detect and sequence any *T. rex* proteins that had miraculously survived inside the bone. And finally, that the device purred and buzzed for an hour before spitting out data describing the molecular contents of the sample.

It was at this moment—when a fragment of 68 million-year-old dinosaur was rendered as strings of letters decipherable only by the most labyrinthine mathematical algorithms—that empirical certainty crumbled. What followed was a complex, contentious, and peculiarly modern scientific argument, one more about software and statistics than bones and pickaxes.

That argument began in earnest in April 2007, when Asara, Schweitzer, and several colleagues announced in the journal *Science* that the mass spectrometer had indeed uncovered seven preserved fragments of protein in MOR 1125. Five of those fragments closely matched sequences of collagen—the most common protein found in bones—from birds, specifically chickens.

The discovery generated international headlines—"Study: *Tyrannosaurus Rex* Basically a Big Chicken"—as the first molecular confirmation of the long-theorized relationship between dinosaurs and birds. It was also the first-ever evidence that protein could survive even a million years, much less 68 million. *The New York Times* reported that the finding "opens the door for the first time to the exploration of molecular-level relationships of ancient, extinct animals." Some news outlets couldn't resist drawing parallels to a certain popular fictional tale. The research, suggested the UK *Guardian*, "also hints at the tantalizing prospect that scientists may one day be able to emulate Jurassic Park by cloning a dinosaur."

Before long, however, a distinctly human subplot emerged. Within 16 months, three separate rebuttals appeared, two in *Science* itself. Many researchers were skeptical of the quality of Asara's data and doubted that collagen could survive so long, even partially intact. "You're talking about something a hundred times older than anything ever sequenced," says Steven Salzberg, director of the Center for Bioinformatics and Computational Biology at the University of Maryland. "If you have extraordinary results, they require extraordinary evidence."

Asara and Schweitzer were forced to parry and retreat, admitting that statistical evidence for one of the protein fragments was too weak for them to claim they'd even found it. The pair's fiercest critic, a UC San Diego computational biologist named Pavel Pevzner, also questioned the other six fragments, demanding that Asara release all of his underlying data. In a caustic 2008 *Science* critique, he compared Asara to a boy who watches a monkey bang away randomly on a typewriter, sees it produce seven words, and "writes a paper called 'My Monkey Can Spell!'" Asara's findings, Pevzner told *The Washington Post*, were "a joke" that would make "serious evolutionary biologists laugh." Then things got contentious.

In many ways, the ongoing case of MOR 1125 exemplifies what can happen when the scientific process—a meticulous consensus built on a foundation of small findings, published in rigorously peer-reviewed journals—is interrupted by a headline-grabbing discovery. As one study catapults into the public sphere, careers and even entire scientific disciplines can come to hinge on its validity. This, then, is a story about what happens when the headlines fade and researchers are left to confirm or debunk the discovery of the week.

The battle over those *T. rex* proteins has spilled out into blogs and conferences, generating a cloud of public accusations—some more founded in science than others. It has also highlighted a real and growing tug-of-war between computational and traditional biological research, with debates that increasingly play out in databases and mathematical formulas. When findings are anchored to digital evidence instead of microscope slides, replicating another biologist's work starts to resemble recalculating a physicist's model. And without the public release of all experimental data, the peer reviews of even the leading scientific journals are rendered meaningless.

As the modern discipline of bioinformatics comes crashing into analog fields like paleontology, researchers are just beginning to grapple with questions that the dinosaur controversy inadvertently unearthed. And in the case of the disputed *T. rex* proteins, the answers may not be as they first appeared.

Pavel Pevzner couldn't care less about dinosaurs. What's important in this *T. rex* business, he tells me one day in his office at UCSD's Center for Algorithmic and Systems Biology, is the thorny mathematical puzzle that arises in the search for proteins. "Biology itself," he says matter-of-factly, "is now a computational science." Pevzner, a 50-something Russian native whose Strangelovian accent morphs his *th*s into *z*s, is tall and rugged, with a perpetual 5 o'clock shadow. He is known as one of the top thinkers in the world of bioinformatics, a man with unquestioned computational chops who views himself as a guardian of statistical rigor. "Pavel is a smart guy, but he kind of has . . . a style," one colleague told me. "He likes to stir the pot." A photo on the university's Web site shows Pevzner in full-on Western gear, complete with 10-gallon hat, a beer in one hand and a rifle in the other.

On this afternoon he is sporting a more typical academic costume of jeans and a blazer. But he seems to be feeling no less the sheriff. "In some areas absolutely fundamental to biology—for example, the sequencing of DNA—there are practically no biologists working in this," he says, only computational scientists. Pevzner specializes in developing algorithms to decode the proteins found in mass spectrometry research. The *T. rex* issue came to him when *Science* asked him to peer-review Asara's paper for publication. Even at first glance, he says, "it was clear that this paper was computationally illiterate."

Following his reasoning requires some understanding of how Asara's protein-detection experiments work. Proteins are chains of amino acids, common molecules known by single-letter names—P for proline, G for glycine, and so on. Schweitzer's biochemical tests on MOR 1125 had hinted that the sample contained amino acids. Asara, then, needed to do three things: detect chains of those amino acids, demonstrate that they were fragments of real proteins, and show that those fragments were organic remnants of the dinosaur itself.

An organism's proteome is the complete set of the proteins it contains. Think of it as a dictionary, a collection of words (proteins) made up of letters (amino acids). Now imagine finding a 68 million-year-old bag that appears to contain thousands of letters strung together in chains of varying lengths. That's MOR 1125. The purpose of mass spectrometry is to spell out those letter strings in order to compare fragments of words against the organism's protein dictionary.

To do that, the letter chains are first split into shorter segments called peptides, which are analyzed to determine their mass. The peptides are then sorted by weight and fragmented to reveal their constituent amino acid sequences, each of which is given a mathematical description called a spectrum. Software-based algorithms then determine the letter sequences of the peptides. There are several respected algorithms available to do this—including Pevzner's—and they can produce somewhat different results.

Once all the letters are identified and placed in sequence, the strings are compared against the dictionaries of different species. Because no *T. rex* proteins had ever been sequenced, Asara had to look for the closest matches in databases of modern animals.

Asara's original paper asserted that the algorithm had identified seven peptides in MOR 1125. The spectra of five of those peptides aligned most closely with chicken collagen, followed by the collagen of frogs and newts. The implication—that *T. rex* was a closer relative of birds than of modern reptiles or amphibians—was just what paleontologists would have predicted.

When the paper landed in Pevzner's inbox, however, it contained the supporting spectra for only those seven peptides. Missing were the tens of thousands of "junk" spectra—strings of letters that Asara's machine had sequenced but couldn't match to anything in the database. Without them, it was impossible to know whether the peptides found in the *T. rex* sample matched chicken peptides out of mere chance. Asara's findings, Pevzner thus asserted, could be nothing more than statistical artifacts—random jumbles of letters that just happened to match words in the dictionary.

Pevzner strongly advised *Science* to reject the *T. rex* findings. But other reviewers—who remain anonymous—disagreed, and the paper was published. As the headlines rolled out, Pevzner expanded on his criticisms in an article of his own. *Science* rejected it.

Over the next year, however, other papers critical of Asara's and Schweitzer's work did appear. Sensing an opening, Pevzner resubmitted his own paper to *Science*, which published it in August 2008. The article lambasted Asara for failing to compute statistical significance values and again demanded that he release the junk spectra. "It is now the turn of the mass spectrometry community," Pevzner concluded, "to question whether the monkey can actually spell."

Meanwhile, the critics carried their attacks into blog postings and comment sections, and then into the press. In some articles, Asara's findings were mentioned alongside an infamous 1994 paper that claimed to have recovered dinosaur DNA, a result later debunked as lab contamination by, among others, Schweitzer. Asara's work—and the entire discovery—appeared increasingly beleaguered. "I knew the reception that this stuff was going to get," Schweitzer says. "I think it's been kind of hard on him."

When Asara refused to release the spectra, he planted himself firmly on one side of a battle over transparency. Scientific journals, as a rule, require that published experimental findings include enough information to allow other researchers to reproduce the results. Traditionally, though, other details can be kept tucked away in lab notebooks, to be mined for further publishable nuggets.

When an experiment relies entirely on statistical data, however, reproducing it in full requires the equivalent of everything in the lab notebook. The oldest branch of bioinformatics, genomics, settled the issue of data disclosure years ago, and today DNA sequencing data is generally released in full when—and sometimes even before—a paper is published. The newer field of proteomics is still a kind of scientific Wild West, but open data advocates argue that publishing the underlying data is just as crucial.

In practice, that ideal runs into the realities of the scientific job market. Researchers depend largely on publication to maintain their funding and academic standing. Releasing mass spec data before scouring it for every potential discovery, Asara complained, would have let others scoop up publishable findings.

To which open-data advocates had a simple answer: tough luck. Much of the research is publicly funded, and the only reason to sit on data is a selfish one.

In the fall of 2008, Asara relented. "I have learned from this process that transparency is always the best policy," he conceded in an online back-and-forth with Pevzner. With that, he posted all 48,216 spectra without restrictions in an online database. "We have nothing to hide," he told me at the time.

Within two weeks, a pair of scientists on the opposite coast turned Asara's own data against him. Martin McIntosh, a proteomics expert at Seattle's Fred Hutchinson Cancer Research Center, and computational biologist Matthew Fitzgibbon downloaded the spectra. When they ran their own set of algorithms, they turned up an unexpected twist: an eighth peptide, one that hadn't appeared in any of Asara's papers. And it yielded a match—not to collagen, but to a hemoglobin peptide found in ostriches.

That finding rang a bell. The pair remembered that Asara's lab had once done a project involving ostrich proteins, giving them an alternate story that could explain Asara's findings: After completing his previous work, they suggested, Asara hadn't managed to scrub all the ostrich molecules out of his equipment. When he then sequenced the *T. rex* sample, he had used some test tube or dropper or machine contaminated with an infinitesimal amount of ostrich protein. Of course the peptides Asara found matched up well with chicken—because they were from another bird.

McIntosh was careful to remain circumspect about the discovery, which, he told me in November, he had submitted for publication: "It just means that there is another parsimonious explanation"—a scientific term for the simplest explanation for a given set of facts. "The positive note is, we couldn't have done this without the data provided by Asara." But he suggested that Asara, in private conversations, was hurting his case by questioning his critics' motives. "We are not trying to get famous on this," McIntosh said. "You know that expression 'dramatic claims require dramatic evidence'?"

I did.

"With a lot of things in science, there is not necessarily anything objective that tells you this is the right answer." It was, he said, more like convincing a jury beyond a reasonable doubt—and here was a piece of evidence casting serious doubt.

On a balmy February afternoon, Pavel Pevzner steps onto a ballroom stage at the San Diego Westin before an audience of fellow scientists at the annual conference of the US Human Proteome Organization. For two years now, he and other critics have been chipping away at the *T. rex* protein discovery. Two researchers even published a paper asserting that the proteins were actually from a bacterial biofilm.

Schweitzer countered that charge convincingly, but it still added to the thick cloud of doubt surrounding the research. In that context, Pevzner's topic—"Mass Spectrometry of T. rex: Treasure Trove of Ancient Proteins or Contamination/Statistical Artifacts?"—has the feel of a final demolition. Pevzner made it clear to me, two weeks prior to taking the stage, that he still thought of Asara's work as "speculative science."

The two researchers, in fact, have been circling each other all day, like kids on a junior high playground. "We're not exactly on a friendly basis," Asara tells me that morning. "But if I see him, I'll say hi, of course." He professes no intention of attending Pevzner's talk or any need to defend himself against whatever computational grenades the Russian is preparing to lob. "The last thing I need is to listen to someone who clearly has a biased view of the data," he had emailed a few weeks earlier.

"I'd like to see him, but I haven't," Pevzner tells me shortly before his talk. It's a strange comment, considering that I just saw him an hour earlier outside a room where Asara was manning a poster presentation of his research. Then, as Pevzner steps up to the stage, Asara slips in and takes a seat.

The T. rex controversy, Pevzner begins, offers "an excuse to discuss the arguably more important topic of statistical significance." He recaps the arguments of his Science article, taking apart the statistical significance of several of Asara's peptides on a giant screen. Asara's original paper, he emphasizes, had contained "no statistical analysis."

A few seats over from me, Asara listens quizzically, one leg propped casually over the other. But at the end of his outstretched arm, a finger nervously taps out a beat on a chair between us.

Two of the peptide identifications, Pevzner says, do look "reasonable," perhaps implying that "there are indeed T. rex collagen peptides in this sample." But then he pulls his trump card: McIntosh and Fitzgibbon's hemoglobin finding, the results of which have not been published but which McIntosh sent to Pevzner. The work yields an alternate hypothesis, Pevzner announces: ostrich contamination— perhaps suggesting that Asara's paper "ought to be withdrawn."

Biology is squishy, Pevzner knows, but numbers are firm, and he believes he's got the computational goods. The hemoglobin can only be from T. rex if you combine the astronomically unlikely possibility that T. rex collagen survived for 68 million years with the equally unlikely survival of hemoglobin. Which raises the question, Pevzner says, of whether "T. rex did indeed taste like chicken. Or maybe like beef?" The crowd chuckles. Asara smiles tightly.

Pevzner concludes that there is a simple choice: "We should either side with Asara et al., and join their claim that they found ostrich hemoglobin peptide from T. rex that was well preserved over 68 million years," he says, "or we should side with Martin's group, who claim it is contamination. Let's take a poll: Who thinks that the hemoglobin is actually T. rex hemoglobin?"

Not a single hand goes up.

Extraordinary claims require extraordinary evidence. Carl Sagan popularized that mantra, and it has served scientific skeptics, and science itself, well. The discovery of 68-million-year-old collagen and hemoglobin fragments in a dinosaur bone is clearly an extraordinary claim. Which leaves us with this question: Who gets to decide what constitutes extraordinary evidence?

Over lunch one day at the conference, I finally sit down with Asara after months of trying to arrange an interview. At 36, he is stocky and pale, with black hair combed straight back into a pile atop his head. Over email he had often sounded besieged and irritated—"if you read our responses, the answer should be quite clear," he curtly replied to my first inquiry about the controversy. In person, however, he is different: open rather than defensive, cheerfully optimistic instead of brusque.

Most of his research—on how cancer cells signal each other—is far removed from dinosaurs, he says. But he concedes that the *T. rex* finding "makes me a name that people recognize." And the evidence he proceeds to lay out for the discovery casts an entirely new light on MOR 1125.

First, he points out that he had used several standard mathematical techniques to reinforce the identification of the collagen peptides in his original paper. Still, Pevzner's original complaint, he says, "made us realize we should be more careful" with computational results. So he asked the author of a different algorithm—one favored for its conservative approach to matching peptides—to rerun the data independently. The results matched the original collagen spectra exactly.

Indeed, in granting the statistical significance of even two peptides, Pevzner was abandoning his original contention—that the proteins were mere statistical artifacts. After all, both criticisms can't be true: If you say the peptides result from contamination, you can't also argue that they're mere ghosts in the numbers. "I think we can reject the army of monkeys scenario," agrees Marshall Bern, a computer scientist at PARC, who like Pevzner writes mass spectrometry algorithms and who ran Asara's fully released data through his own algorithm.

Pevzner, when I call him later, concedes as much. "After the spectra were released, it became clear that at least two are reasonable quality spectra," he says. "The new argument came in, and this is contamination."

So sample MOR 1125 unequivocally contains some proteins. But are they from a *T. rex* or from an ostrich? For starters, Asara says, the hemoglobin peptide matches more than 30 birds, which suggests that McIntosh picked ostrich because he knew of Asara's previous work with that species.

What's more, Asara conducted his ostrich and *T. rex* experiments a year and a half apart, separated by roughly 1,500 mass spectrometry runs. According to Asara, none of those spectra, nor samples of the soil surrounding the fossils, nor his daily control runs—in which he sequences known solutions to check for contaminants—turned up any ostrich hemoglobin. Also, the ostrich that Asara had sequenced hadn't even produced the particular hemoglobin sequence McIntosh matched. And *Science* had actually rejected McIntosh's ostrich paper after receiving Asara's response.

Schweitzer, meanwhile, had published several articles reporting evidence of collagen in MOR 1125 obtained using traditional biological techniques. That work had been done in her own lab, on samples never sent to Asara. The pair also collaborated on an identical study of several-hundred-thousand-year-old mastodon proteins—without contamination or criticism.

When I ask McIntosh after the conference how he explains away this evidence, he says, "It's routine that you run a bunch of samples and only one of them is contaminated." The burden of proof lies with Asara, he contends. McIntosh maintains, too, that a certain chemical modification on the hemoglobin makes it more likely to be contamination (to which, of course, Asara offers a rebuttal).

Over lunch, Asara asserts that between his work and Schweitzer's, they have answered the critics. "This is biology we're doing here; it's not just computational analysis," he concludes, between bites of a BLT. "This is a story about protein preservation. When you look at all the validations we did, how can we make the story more convincing?"

Well, there is one way. In early May, a new paper by Asara and Schweitzer—together with more than a dozen coauthors—appeared in *Science*. In it, the team has replicated their protein experiments on MOR 2598, a bone fragment from an 80-million-year-old hadrosaur, an entirely different species, dug up in a different part of Montana in 2007.

This time, they have used even more rigorous controls, handling the fossils with sterile instruments from the beginning of the excavation. They have replicated both Schweitzer's biochemical results (which show evidence of degraded cells and blood vessels) and Asara's mass spec data (which reveal eight collagen peptides) in independent labs. Asara himself used a mass spec machine with much higher resolution and adhered to Pevzner's demands for rigorous statistical analysis. Once again, the ancient protein fragments have lined up with bird collagen. But they lined up most closely with something else: the *T. rex* peptides reported two years ago by Asara.

McIntosh declares himself swayed, though still circumspect. "It's a nice bit of work," he tells me. "I think they've been doing a good job of shutting the door. Whether the door is truly locked or not, I don't know." Some other explanation could potentially win out over time. But the hemoglobin-based ostrich contamination hypothesis, he says, "doesn't really bear on what they're trying to prove here."

Pevzner, characteristically, is still playing the sheriff. "I'm glad that Asara called the previous criticism appropriate," he says. "I had a commentary that their analysis was unprofessional; they agreed with this. I had a commentary that this work couldn't be evaluated unless they release the data; they agreed with that."

He maintains that Asara and his colleagues have erected a "wall of silence" around the issue of McIntosh's hemoglobin peptide discovery, which goes unmentioned in the new paper. "This is much bigger news than the collagen," he says. And the researchers are keeping it quiet, he adds, precisely because it is so extraordinary as to cast doubt on their conclusions.

It's a bold claim, but one that McIntosh himself swats down. Since the hemoglobin finding was not published, he points out, it essentially remains a scientific rumor—not a solid theory that demands addressing. Now, to be convincing, Asara's critics are the ones who need evidence to back their alternate hypotheses. "It's up to them to demonstrate it," McIntosh says.

Asara and Schweitzer, in other words, have done just what the critics asked. They've built a rigorous scientific case for the survival of 68-million-year-old proteins from a beast that animates children's imaginations. If it continues to hold up, it is research worthy of its international fanfare. The slow, grinding process of science, freed from the headlines, is working just as it's supposed to.

The one lesson that all sides of the debate now agree on is that the new age of computational biology must be one of data transparency. Such disputes can only be resolved—and the scientific method can only survive the digital age—if scientists dump their digital notebooks online for anyone to try to replicate. And in that sense, Pevzner has been right from the beginning.

Indeed, when *Science* published the new paper in early May—the one that Asara knew would silence many of his critics—he made a special arrangement to release the entire data set online the same day. Extraordinary claims, as they say, require extraordinary evidence.

4

Mass Extinction: What Killed the Dinosaurs?

From the Smithsonian Scientific Series (1938), public domain

Many scientists believe that a meteor impact, or a series of such impacts, led to the extinction of the dinosaurs.

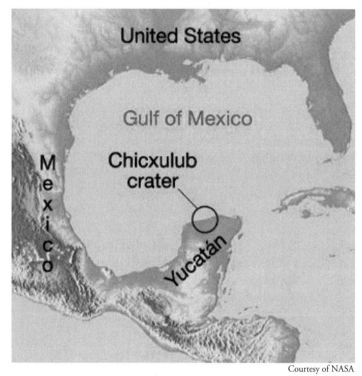

Did the meteor strike that created the massive Chicxulub Crater 65 million years ago also cause the extinction of the dinosaurs?

Editor's Introduction

Perhaps nowhere in the field of dinosaur study does the pendulum of debate swing wider and more vigorously than on the subject of extinction. Many species of animals have risen and fallen over the course of Earth's history, but few fascinate us like these mysterious creatures. Whether roaring on the big screen or towering over us in natural history museums, dinosaurs have an air of invincibility. It's difficult to imagine what could have wiped them out.

The chapter begins with a pair of *Natural History* articles that argue both sides of the question, "Were Dinosaurs the Victims of a Single Catastrophe?" In the first, University of Rhode Island geosciences professor David Fastovky answers in the affirmative, blaming a lone asteroid strike. As he sees it, extinction was "geologically instantaneous," meaning it took anywhere from a minute to many thousands of years. After the impact, he writes, small, cold-blooded, aquatic animals had the best chance for survival, as dust and other debris in the atmosphere blocked sunlight and eliminated much of the food supply. In "No, It Only Finished Them Off," San Diego State University biology professor J. David Archibald makes the counterargument. While he doesn't deny the occurrence of asteroid impacts, he claims most dinosaur species had already disappeared by the time of these events.

In the next selection, "New Theory on Dinosaurs: Multiple Meteorites Did Them In," *New York Times* writer William J. Broad discusses a pair of craters—one in the Ukraine, the other at the bottom of the North Sea—now thought to have been formed 65 million years ago. The craters lend credence to the idea that the Earth was struck by a series of objects, rather than the single asteroid Fastovsky and numerous other scientists believe spelled doomsday for the dinosaurs.

Most lone-asteroid theorists point to a crater on the Yucatan Peninsula in Mexico, and while scientists agree this was the scene of a major impact, there remains some question regarding when it occurred. As Angela Botzer reports in "Yucatan Asteroid Didn't Kill the Dinosaurs, Study Says," the crater's age is still up for debate. Recent core samples and fossil discoveries have led Princeton University geosciences professor Gerta Keller to conclude that the asteroid hit Mexico 300,000 years before the dinosaurs died off. Keller believes asteroid impacts were just one of three ingredients—"massive volcanic eruptions" and subsequent climate change being the other two—responsible for the extinction.

Botzer's article makes mention of "Shiva," a massive crater discovered off the coast of India. In the next article, "I Am Become Death, Destroyer of Worlds," an *Economist* writer outlines a new theory proposed by Sankar Chatterjee, the Texas Tech professor who named the Indian crater for the Hindu god of destruction and renewal. Chatterjee believes the Shiva asteroid—40 kilometers in diameter—landed in an active volcano zone, finishing off the dinosaurs and forming the islands that now make up the Republic of Seychelles. If Chatterjee's hypothesis is correct, the dinosaurs were the victims of what *The Economist* deems "atrocious bad luck," as they suffered two catastrophic asteroid impacts during a 300,000-year time span marked by intense volcanic activity.

The chapter ends with "A Theory Set in Stone: An Asteroid Killed the Dinosaurs, After All," in which Katherine Harmon reports on an exhaustive review conducted by 41 dinosaur experts. Drawing on 30 years of research papers and data from 350 different sites, the scientists upheld the single asteroid hypothesis, dismissing those based on global warming or multiple strikes.

Were Dinosaurs the Victims of a Single Catastrophe?*

Yes, and an Asteroid Did the Deed

By David Fastovsky
Natural History, May 2005

The idea that a single, spectacular, catastrophic event—an asteroid impact—at the end of the Cretaceous period, 65 million years ago, obliterated all the nonbird dinosaurs (as well as many other organisms) is a simple, attractive scenario. But is it more accurately described as *simplistic*? Since not all life was wiped off our planet, there must have been winners as well as losers in the Cretaceous endgame. Surely survival occurred for better reasons than a mere roll of the cosmic dice!

Yet in the past fifteen years it has become clear that the extinction of the dinosaurs was geologically instantaneous. Geological instantaneity, however, is an inexact quantity. From a vantage point of 65 million years after the event, paleontologists cannot resolve time spans of less than tens of thousands of years: whether the extinction took a minute or many thousands of years may never be known. Still, 10,000-year timescales rule out events that lasted millions of years, and those include a whole class of gradual, earthbound processes.

In three separate studies in western North America (the only place where these issues have been studied), the diversity of dinosaur fossils was carefully recorded, meter by meter, through rocks that record the Cretaceous-Tertiary (K/T) boundary. In each case, paleontologists failed to identify any decrease in dinosaur diversity in the 2 million years or so preceding the boundary. In fact, every published, quantitative, field-based, stratigraphically refined study addressing this question has concluded that dinosaur diversity was unchanged up to the K/T boundary: the final extinction was thus geologically instantaneous.

When it comes to the ecology of survival, paleontologists are in refreshing agreement: your chances of surviving were pretty good if you were small and cold-blooded (correctly termed ectothermic). But your best bet was to be aquatic.

At first blush, those attributes might not seem the ideal armament against incoming asteroids. But they do seem to have been keys to survival. Although the exact effects of large-body impacts on Earth remain uncertain, there is a general consensus that such impacts probably have two kinds of dire consequences: dust, smoke, and debris in the atmosphere blocking sunlight for several months, and an instantaneous pulse of thermal energy igniting global fires.

For both those effects, being small, ectothermic, and aquatic may have been the secret to survival. For as long as sunlight was blocked, photosynthesis would have ceased, reducing much of Earth's available foodstuffs to detritus. Dinosaurs and other organisms dependent on "primary production"—fresh plants and meat—would have become effectively helpless. But aquatic animals, which tend to feed on detritus, may have been protected. On land, many Cretaceous mammals were likely part of detritus-based food chains. Furthermore, many were small enough to have lived in burrows (as many small mammals do today), and so they could have been protected from the thermal pulse. Finally, in the face of a global heat pulse and fires, small size and aquatic refuge offered nearly ideal shelter in what had become an inhospitable world. In short, the ecology of both winners and losers reflects the imprint of the impact with surprising fidelity.

Were Dinosaurs the Victims of a Single Catastrophe?*

No, It Only Finished Them Off

By J. David Archibald
Natural History, May 2005

Some 65 million years ago, Murphy's Law applied—almost everything that could have gone wrong did: A huge bolide, or asteroid, struck Earth. Globally, the seas receded. Fissures on the Indian subcontinent spewed forth thousands of cubic kilometers of material. All three events took place in rapid succession, toward the end of the Cretaceous period. Each of them is thought to have been the largest event of its kind in the past 250 million years, and each is thought to have played a role in the demise of the nonbird dinosaurs. Each event left obvious physical and chemical proof of its occurrence in the rock record. That much is clear. But how can paleontologists measure the effects of such events on the creatures living at that time?

The most powerful method is simply to read, in the fossil record, which animals survived and which did not. Only western North America, though, preserves a reasonably continuous fossil record of the land and freshwater vertebrates for the last 10 million years of the Cretaceous and on through the Cretaceous-Tertiary (K/T) boundary. In those last 10 million years of the Cretaceous, but well before the K/T-boundary events, the most recent compilations show an unequivocal decline in the diversity of dinosaur species. In fact, before the time of the boundary is reached, between one-third and one-half of all dinosaur species—mostly such relatively common groups as the duck-billed and horned dinosaurs—had already disappeared.

The analysis of the final million years of the Cretaceous is more problematic, because the precision required is far greater than is discernible in the fossil record. A recent study in North Dakota noted little or no change in the vertebrate fauna throughout the thickness of the Hell Creek Formation. Those data were cited to argue that a bolide impact must have suddenly terminated the nonbird dinosaurs at the top of this formation.

Yet in the uppermost five meters of the formation only two dinosaurs could be identified well enough to specify their generic name. What happened to the other eighteen or so nonbird dinosaur species present in the Hell Creek Formation? No one knows whether they survived to the time of the boundary or became extinct thousands of years before it.

Apart from the problems of detecting rates of dinosaur extinction, we can examine the pattern of total vertebrate extinction. Of 107 species of vertebrates known from Hell Creek, about half had disappeared by the time corresponding to the K/T boundary. Of those extinctions, 75 percent are concentrated in just four groups: lizards, marsupials, sharks (and their relatives), and nonbird dinosaurs. The lizards may have faced habitat loss from increasing rainfall in the Hell Creek region near the end of the Cretaceous. As sea levels fell, the Bering land bridge enabled the precursors of modern hoofed mammals to enter North America and outcompete other mammals, notably the marsupials. The sharks, too, lost their habitat as the seas retreated. And the nonbird dinosaurs? With the loss of inland seas, the low coastal plains, from which almost all of the fossils of these animals are known, shrank and fragmented.

Then the bolide struck Earth. Many consequences of this impact have been proposed—global wildfire, extended periods of darkness, sharp temperature increases, tsunamis, and hurricanes. Other suggested effects—notably acid rain and a sharp drop in the temperature—now seem extremely unlikely, given the fossil record. The most recent proposed consequence has been sudden infrared heating. That might explain why large creatures such as dinosaurs died, whereas smaller species survived by taking refuge in holes, crevices, or under a thin layer of water.

Whatever the results of the impact, though, it only finished a job that earthbound factors had already begun. The dinosaurs and other vertebrate species had already become vulnerable to extinction.

New Theory On Dinosaurs*

Multiple Meteorites Did Them In

By William J. Broad
The New York Times, November 5, 2002

For more than a decade, most scientists have believed that the extinction of the dinosaurs was caused by a single event: the crash of an immense body from outer space, its explosive force like a hundred million hydrogen bombs, igniting firestorms and shrouding the earth in a dense cloud of dust that blocked sunlight and sent worldwide temperatures plummeting.

The theory gained wide acceptance in 1991, after the discovery of a crater buried under the tip of the Yucatan Peninsula. The giant gash stretched 110 miles from rim to rim, and its age was found to be 65 million years, the same time as the death of the dinosaurs.

Now, however, scientists working in Ukraine have discovered that a well-known but smaller crater, some 15 miles wide, had been inaccurately dated and is actually 65 million years old, making the blast that created it a likely contributor to the end of the dinosaurs.

So too, a British team has recently found a crater at the bottom of the North Sea dating to the same era and stretching over 12 miles in a series of concentric rings.

The discoveries are giving new support to the idea that killer objects from outer space may have sometimes arrived in pairs or even swarms, perhaps explaining why the extinctions seen in the fossil record can be messy affairs, with species reeling before a final punch finishes them off.

"It's so clear," said Dr. Gerta Keller, a geologist and paleontologist at Princeton, who studies the links between cosmic bombardments and life upheavals. "A tremendous amount of new data has been accumulated over the past few years that points in the direction of multiple impacts."

But Dr. Keller added that many scholars had staked their reputations on the idea of a single dinosaur-ending disaster and were reluctant to consider the new evidence. "Old ideas," she said, "die hard."

Her own research, Dr. Keller added, suggests the reality of multiple strikes and raises doubts that the Yucatan rock, whose crater is known as Chicxulub, was the event that sealed the dinosaurs' fate. Instead, she said, the main killer "has yet to be found."

The ferment is prompting scientists around the globe to look for new craters and to reassess the ages of old ones in search of clues to the wave of global extinction that did in thousands of species—not only the dinosaurs but many plants, fish and plankton—at the end of the Cretaceous period.

"There are over 170 confirmed craters on earth and we know the precise impact age of only around half," said Dr. Simon P. Kelley of the Open University in Britain, who found the dating error on the Ukraine crater, along with Dr. Eugene P. Gurov of the Institute of Geological Sciences in Ukraine. Even in the United States, he added, several craters are poorly dated.

"In the U.K., we have a phrase, 'You wait an hour for a bus, then three come along all at once,'" he remarked in an interview. "Maybe impacts are like that."

The idea that a giant intruder from outer space killed off the dinosaurs was proposed in 1980 by Dr. Luis W. Alvarez; his son, Dr. Walter Alvarez; and their colleagues at the University of California at Berkeley. It was met with great skepticism at first, but in time became the standard belief.

In his 1997 book, "T. Rex and the Crater of Doom," Dr. Walter Alvarez, a geologist, said he had considered the possibility of multiple impacts until 1991 and the discovery of the huge Yucatan crater, which seemed big enough to solve the mystery on its own.

Dr. Kelley and Dr. Gurov presented their findings from Ukraine in the August issue of the journal *Meteoritics & Planetary Science*. In geologic time, the twin birth throes of the Ukraine and Yucatan craters, they note, suggest rather than prove "that they combined to lead to the mass extinction" at the end of the Cretaceous period and raise questions of other possible cosmic killers.

Known as Boltysh, the newly dated crater lies in eastern Ukraine in the basin of the Tyasmin River, a tributary of the Dnieper. Though just 15 miles wide, the buried crater, whose presence is revealed by deep jumbled masses of melted and broken rocks, is surrounded by a ring of rocky debris that extends over many hundreds of square miles, conjuring up a fiery cataclysm. The two scientists say in their report that this kind of crash today would have devastated a densely populated nation.

Over the years, scientists had analyzed rocky samples from the Boltysh crater and found ages ranging from 88 million to 105 million years.

The new dating of the crater by Dr. Kelley and Dr. Gurov used a highly accurate method that carefully measures the ratio of two isotopes of the element argon, a colorless, odorless gas that makes up about 1 percent of the earth's atmosphere. Argon-argon dating works because the isotopes decay at different rates. By measuring the ratio, it is possible to estimate how long ago the sample melted to trap atmospheric argon.

Dr. Kelley and Dr. Gurov report that seven samples of melted rock from the depths of the Boltysh crater yielded an average age of 65.2 million years, with an accuracy of plus or minus 600,000 years.

By contrast, Chicxulub (pronounced CHEEK-soo-loob) has been dated to 65.5 million years, plus or minus 600,000. Given the range of dating uncertainty, the two impacts that made the craters may have occurred simultaneously or been separated by thousands of years.

Scientists have recently looked more favorably at the idea that comets can travel in packs. In the 1980's, a few speculated that comet showers might produce strikes on the earth over a period of a million years or so to bring on extinctions. The idea gained support in 1994 when the comet Shoemaker-Levy 9 was fractured by the gravitational pull of Jupiter into 21 discernible pieces that then, one by one, bombarded the planet.

Dr. Kelley and colleagues at the University of Chicago and the University of New Brunswick, writing in the journal *Nature* in 1998, gave precise dating evidence to argue that a similar kind of celestial barrage hit the earth 214 million years ago. Spread over Europe and North America, the chain of five craters, they wrote, indicated that a large comet or asteroid had broken up and struck the earth in a synchronized assault.

Today, Dr. Kelley said, the odds of the Boltysh and Chicxulub craters' having formed simultaneously, like the chain, are not great. Still, even if their times of impact prove to have been only close, experts say, the one-two punch could still have added to the global turmoil that did in the dinosaurs and other creatures.

Beneath the North Sea, two British oil geologists have found another crater, buried under hundreds of feet of ooze, that may have contributed to the chaos. Writing in the Aug. 1 issue of *Nature*, Simon A. Stewart and Philip J. Allen said they were able to date the 12-mile structure to a period 60 million to 65 million years ago. They named it Silverpit, after a nearby sea-floor channel.

Experts say the new finds may answer an old criticism of the single-impact theory. Critics, especially the paleontologists who specialize in dinosaur extinction rates, had long noted that the fossil record of the late Cretaceous shows a slow decline of many life forms rather than a single vast die-off. That seemed inconsistent with a cosmic catastrophe.

But now, the emerging family ties among the Boltysh, Silverpit and Chicxulub craters suggest that a series of impacts may have driven or contributed to this slow decline.

Dr. Keller of Princeton and her colleagues have found signs of other intruders from outer space that hit at slightly different times about 65 million years ago, strengthening the gradualist idea.

Working in northeastern Mexico, they discovered that glass spheres of melted rock once thought to have been thrown out by the Chicxulub impactor were more likely the result of at least two separate disasters, about 300,000 years apart. They recently presented their findings in a paper for the Geological Society of America.

Moreover, Dr. Keller said, the evidence suggests that the earlier of the two cataclysms formed the Chicxulub crater, making its arrival too early to account for the killer punch of the dinosaur extinction.

Geologic clues that she and her colleagues are collecting from Mexico, Guatemala, Haiti and Belize, Dr. Keller said, suggest that a barrage of cosmic bodies hit the earth over the course of 400,000 years. The first was the Chicxulub event, the second the unlocated impactor at the end of the Cretaceous period and then a straggler some 100,000 years later.

Strong evidence exists for three impacts at the end of the Cretaceous era, Dr. Keller said, followed by wide climate shifts that lasted through the turbulent period.

While geologists hunt for other craters and impact events, they say the most compelling evidence of all may have vanished. Since the earth's surface is more than 70 percent water, it is likely that most signs of speeding rocks from space disappeared long ago in the churning geological processes that constantly renew the seabed. The North Sea, being relatively shallow, is an exception.

Despite the inherent difficulties of the research, Dr. Kelley of the Open University said he planned to redouble his hunt to "try to solve this problem."

Yucatan Asteroid Didn't Kill the Dinosaurs, Study Says[*]

By Angela Botzer
National Geographic News, March 9, 2004

Sixty-five million years ago, a city-size asteroid slammed into what is now Mexico's Yucatán Peninsula at a site known as Chicxulub (pronounced CHEEK-shoo-loob). The impact left a crater 110 to 170 miles (180 to 280 kilometers) wide and spewed massive volumes of ash into Earth's atmosphere.

Did that cataclysmic event trigger the extinction of the dinosaurs and 70 percent of the world's other species? For over a decade, most scientists said yes.

But authors of a controversial new study published in the *Proceedings of the National Academy of Sciences* (online edition) contend that the asteroid behind the Chicxulub crater impacted Earth 300,000 years earlier than previously thought. They say a second, as yet unidentified asteroid impact must have caused the mass extinction popularly attributed to the Chicxulub asteroid.

Princeton University professor of geosciences Gerta Keller led the study, which analyzed new core samples drilled at Chicxulub. The drilling was "done with the express purpose to solve the ongoing controversy of what killed the dinosaurs and prove once and for all that this is the impact that caused the mass extinction," Keller said.

However, Keller said close examination of layers in the core samples shows that the prevailing theory that the Chicxulub asteroid killed the dinosaurs "seems to be wrong."

"The Chicxulub impact hit Yucatán about 300,000 years before the mass extinction. Another impact occurred at the time of the mass extinction," she said.

While asteroid impacts played a role, Keller says several hundred thousand years of "massive volcanic eruptions" contributed to climatic changes that precipitated the mass extinctions that marked the end of the Cretaceous period.

TRACKING THE SMOKING GUN

The single-asteroid-impact extinction theory first appeared in 1980. At the time father and son scientists Luis and Walter Alvarez produced a daring hypothesis: An asteroid or meteorite may have struck the Earth, triggering mass extinctions that drew the curtain on Cretaceous period.

The pair based their theory on research which found that the K-T (Cretaceous-Tertiary) boundary—a geologic layer present in rock formations around the world—exhibits a thin layer of clay rich in iridium. Rare on the Earth's surface, the element is more commonly found in extraterrestrial sources like asteroids, or deep inside the Earth's core.

That iridium-rich layer the Alvarezes described was later found in multiple K-T localities around the world. The race was on to find the smoking gun: the crater impact site itself.

Alan Hildebrand, an associate professor at the University of Calgary, Canada, took the lead in 1990, discovering the impact crater in Yucatán. The crater appeared to be a probable source for the iridium and seemed to be the impact site of the asteroid that destroyed the dinosaurs. Using boreholes and geophysical evidence, scientists found the crater buried under sediments dating to the Tertiary period, the era between 65 and 1.8 million years ago, which followed the Cretaceous.

FOSSIL EVIDENCE

But Keller and her colleagues say their research proves otherwise.

Keller has studied the Chicxulub site and other impact-crater sites around the world for the past decade. She believes that the asteroid impact behind Chicxulub coincided with a "time of massive volcanism, which led to greenhouse warming."

Keller says those three events—the Chicxulub asteroid impact, volcanism, and climate change—"led to high biotic stress and caused the decline of many tropical species populations," but not mass extinctions. That die-off didn't occur until later. However, Keller does believe that the initial confluence of volcanic activity, global warming, and the Chicxulub asteroid impact ultimately contributed to the mass extinction.

Key to Keller's assertions is a 20-inch-thick (50-centimeter-thick) layer of limestone found between the K-T boundary and the impact breccia, or molten lava and rocky debris, laid down when the Chicxulub asteroid collided with Earth.

Keller and her colleagues believe that the thickness of the limestone layer—a type of sedimentary rock characteristically formed under large bodies of water like oceans, seas, and lakes—indicates that it accumulated in the crater over some 300,000 years after the impact. As proof, Keller points to fossils of microscopic organisms called foraminifera and fossil burrows present in the limestone layer.

According to Keller, those fossils indicate the sediment was deposited after the asteroid impact but before the period of mass extinction that marked the end of the Cretaceous.

Many other scientists disagree with that interpretation, however. They say the layer of fossil-rich limestone was deposited quickly as backwash and infill caused by a huge tsunami that followed the Chicxulub asteroid's impact with Earth. The layer, they say, did not take 300,000 years to accumulate.

In her defense, Keller says the quick-accumulation theory is unsupported by evidence that would have been found during her analysis of core samples gathered at Chicxulub and 45 localities in northeast Mexico.

But Alan Hildebrand, a proponent of the quick-accumulation theory, says the burrows were "made by organisms digging after the fireball layer was deposited."

Thomas R. Holtz, Jr., a vertebrate paleontologist at the University of Maryland in College Park, supports the view that the limestone was quickly laid down as crater infill. He said he is not surprised that Cretaceous fossils were found in the limestone layer.

"If an asteroid clobbered the Eastern seaboard of the U.S. today, I would expect that most of the infilling would be Chevys and Hondas and shopping malls and houses and cows and McDonald's burger wrappers," Holtz said. "Only a tiny bit might be mastodons and Clovis points and Miocene whales." In other words, the crater would quickly fill with objects common on Earth at the time of impact.

So where do researchers in the Keller camp look next for the possible K-T crater? Keller says she's unsure, although "some scientists have suggested it could be a structure called Shiva, in India. We have no convincing evidence so far that this is the case."

I Am Become Death, Destroyer Of Worlds[*]

The Story of How the Dinosaurs Disappeared Is Getting More and More Complicated

The Economist, October 2009

Everyone knows that the dinosaurs were exterminated when an asteroid hit what is now Mexico about 65m years ago. The crater is there. It is 180km (110 miles) in diameter. It was formed in a 100m-megatonne explosion by an object about 10km across. The ejecta from the impact are found all over the world. The potassium-argon radioactive dating method shows the crater was created within a gnat's whisker of the extinction. Calculations suggest that the "nuclear winter" from the impact would have lasted years. Plants would have stopped photosynthesising. Animals would have starved to death. Case closed.

Well, it now seems possible that everyone was wrong. The Chicxulub crater, as it is known, may have been a mere aperitif. According to Sankar Chatterjee of Texas Tech University, the main course was served later. Dr Chatterjee has found a bigger crater—much bigger—in India. His is 500km across. The explosion that caused it may have been 100 times the size of the one that created Chicxulub. He calls it Shiva, after the Indian deity of destruction.

Dr Chatterjee presented his latest findings on Shiva to the annual meeting of the Geological Society of America in Portland, Oregon, on October 18th. He makes a compelling case, identifying an underwater mountain called Bombay High, off the coast of Mumbai, that formed right at the time of the dinosaur extinction. This mountain measures five kilometres from sea bed to peak, and is surrounded by Shiva's crater rim. Dr Chatterjee's analysis shows that it formed from a sudden upwelling of magma that destroyed the Earth's crust in the area and pushed the mountain upwards in a hurry. He argues that no force other than the rebound from an impact could have produced this kind of vertical uplift so quickly. And the blow that caused it would surely have been powerful enough to smash ecosystems around the world.

DOUBLE WHAMMY

In truth, agreement on the cause of the mass extinction at the end of the Cretaceous (when not only the dinosaurs, but also a host of other species died) has never been as cut and dried among palaeontologists as it may have appeared to the public. One confounding factor is that the late Cretaceous was also a period of great volcanic activity. In India, which was then an island continent like Australia is today (it did not collide with Asia until 50m years ago), huge eruptions created fields of basalt called the Deccan Traps. Before the discovery of Chicxulub, the climate-changing effects of these eruptions had been put forward as an explanation for the death of the dinosaurs. After its discovery, some argued that even if the eruptions did not cause the extinction, they weakened the biosphere and made it particularly vulnerable to the Chicxulub hammer-blow.

There are also puzzling anomalies in the pattern of extinction. The greatest of these is that, as fiery and horrible as the impact would have been, the survivors included many seemingly sensitive animals like birds, frogs and turtles. Moreover, close inspection of the fossil record shows that many "Cretaceous" species disappear both well before, and well after, the signs of the impact that are found in the rocks.

Ironically, it was while he was investigating the Deccan Traps that Dr Chatterjee came across the evidence for Shiva. First, he found dinosaur nests that had been built between lava flows 10–15 metres thick—evidence that the animals were coping well with the volcanic activity rather than being weakened by it. Then, quite suddenly, 65m years ago, a layer of lava nearly 2km thick appears. This led him to wonder what could possibly have caused such a sudden volcanic surge.

He knew that the west coast of India had been the site of an ancient impact of unknown age and size. It was not until he was reading through a paper published by an oil company that had collected geological information in the area that he realised the volcanic surge he had seen might be related to a cosmic collision.

Further examination revealed a crater rich in shocked quartz and iridium, minerals that are commonly found at impact sites. (These are also the telltales in distant layers of ejecta that the rock in question has come from an impact.) Most important, the rocks above and below Shiva date it to 65m years ago. Dr Chatterjee therefore suggests that an object 40km in diameter hit the Earth off the coast of India and forced vast quantities of lava out of the Deccan Traps. As well as killing the dinosaurs the impact was, he proposes, responsible for breaking the Seychelles away from India. These islands and their surrounding seabed have long looked anomalous. They are made of continental rather than oceanic rock, and seem to be a small part of the jigsaw puzzle of continental drift rather than genuine oceanic islands.

This story, though, raises the question of why there is but a single ejecta layer of iridium and shocked quartz in late Cretaceous rocks around the world. One answer might be that the two impacts were, in effect, simultaneous—that the objects which created Shiva and Chicxulub were the daughters of a comet that had broken up in space and hit the Earth a few hours apart, as the pieces of comet Shoemaker-Levy 9 hit Jupiter in 1994. Other, smaller craters in the North Sea and Ukraine have been prayed in aid of this theory. Recent research suggests that there was, in fact, no impact in the North Sea at all, but the Ukrainian site does appear to be the result of a genuine collision that took place at the time of the dinosaur extinction and could therefore be connected with Shiva. But even if it is, it is so much smaller that its environmental effect would have been both minor and local.

Extensive dating research at Chicxulub, however, now suggests that the object which created that crater actually struck 300,000 years earlier than the dinosaur extinction, meaning there really should be two ejecta layers. That there are not could be explained by the fact that the accumulation of sediment in most rocks is so slow that the two layers are, in effect, superimposed. Alternatively, it could be that no one has been looking for two layers, so they have not seen the double signature or have ignored its significance. Indeed, two iridium layers have been found in some places. Anjar, an Indian town north of the impact site, is one. That is leading Dr Chatterjee to suggest that the two big impacts did take place at different times.

The picture that is emerging, then, is of a strange set of coincidences. First, two of the biggest impacts in history happened within 300,000 years of each other—a geological eyeblink. Second, they coincided with one of the largest periods of vulcanicity in the past billion years. Third, one of them just happened to strike where these volcanoes were active. Or, to put it another way, what really killed the dinosaurs was a string of the most atrocious bad luck.

A Theory Set in Stone[*]

An Asteroid Killed the Dinosaurs, After All

By Katherine Harmon
Scientific American, March 4, 2010

Although any *T. Rex*–enthralled kid will tell you that a gigantic asteroid wiped the dinosaurs off the planet, scientists have always regarded this impact theory as a hypothesis subject to revision based on further evidence gathered from around the globe. Other possible causes, such as volcanism and smaller, multiple asteroid strikes, never actually went away, and over the years researchers raised important points that did not fully jibe with a history-changing celestial impact near the Yucatan peninsula one awful day some 65.5 million years ago.

A group of 41 researchers have pored over the evidence and decided that—in accordance with the original postulate put forth 30 years ago by a team led by father and son researchers Luis and Walter Alvarez—it was, indeed, a massive asteroid that slammed into Earth, creating Chicxulub Crater on Mexico's Gulf Coast, that killed off many of the species on the planet, including the non-avian dinosaurs.

The review, published online March 4 in *Science*, evaluated the whole picture, according to Kirk Johnson of the Research and Collections Division at the Denver Museum of Nature and Science and co-author of the paper. And that meant assessing the other theories that have been put forth about what spelled death for the dinosaurs.

FIERY FAILURES

The researchers dismiss the theory that the volcanism that produced the great Deccan Trap formation in western India at the end of the Cretaceous period might have produced enough sulfur and carbon dioxide to initiate a massive shift in climate. They note that pinpointing the times when the heavy volcanism occurred is sketchy, and it likely kicked off some 400,000 years before the extinction event. In fact, as Johnson noted in a March 3 conference call with reporters, the emissions from these volcanoes likely warmed the planet slightly, actually making life easier for many animals and encouraging diversification and dispersion over wider geographical areas.

Some scientists have pointed to multiple layers of impact residue as evidence that there was more than one asteroid involved in generating the extinction. This theory did not seem to measure up, either. Johnson says they see "no evidence for multiple impacts," and sites that had turned up these various layers were so close to Chicxulub itself that the chaotic event likely churned the layers into different locations in the sediment.

An assertion that the impact occurred hundreds of thousands of years before the extinctions also failed to hold water with the researchers. Evidence of Cretaceous period shells on top of the impact crater are likely not a sign that the animals persisted after the impact, but rather that they got "washed into the hole," Johnson noted.

GLOBAL GROUND ZERO

The researchers assessed reports from some 350 sites all over the globe that had evidence of the impact—whether it was a dusting of iridium (an element much more common in extraterrestrial objects) or bits of shocked quartz—and could be traced back to the Chicxulub location. In some areas near the crater, the layer was 80 meters thick, pointing to one single devastating day for life on the planet.

"That's the single best explanation for the extinction of so many groups," says Neil Landman, a curator at the American Museum of Natural History in New York City and was not involved in the review, about the single impact theory.

"We've examined sites around the world," he notes of his study of ammonoids, which are shelled cephalopods that went extinct after the Cretaceous. And from the work he and his colleagues have done, he says, the evidence for the Chicxulub asteroid impact is the most consistent. "I'm very comfortable with this explanation."

A MASSIVE BLOW

Based on the size of material from rocky shrapnel and the crater diameter, researchers have estimated the dino-demolishing object to be some 10 kilometers across. And when it struck—at about 20 kilometers per second—it created an instant crater about 100 kilometers wide and 25 to 30 kilometers deep "almost piercing the crust of the Earth," Johnson noted. The final crater that formed after the initial impact was about 180 kilometers across and two kilometers deep, which is still close to the depth of the Grand Canyon, Johnson pointed out.

The impact spewed rock so high, some of it likely was shot into orbit, whereas other pieces entered the upper atmosphere, reheating as they fell back to the ground. The jolt would have spurred massive earthquakes—some surpassing magnitude 11—tsunamis and landslides. While examining ammonoid fossils in southeastern Missouri, Landman says, he found a shallow water site that was "just immediately covered over by a jumble of stuff," he says. "I think what we're seeing is a tsunami," which might have reached as far from the Yucatan impact site as southern Illinois.

Perhaps most devastating, however, the crash would have caused acid rain and darkness, as particulate matter blocked sunlight, prohibiting photosynthesis in both land and water ecosystems, effectively shutting down large swaths of the food chain. Directly after the extinction event, ferns (which reproduce from spores) proliferated and species that depended on detritus seemed to survive.

From Landman's study of ammonoids, he points out that even for groups that eventually went extinct after the collision (producing the so-called K–T boundary in the fossil record), the asteroid's impact did not mean sudden eradication. "There seems to be some suggestion of some survival for awhile after the event," he says. Fossils found above the iridium layer show that ammonoids might have survived "for tens to possibly hundreds of years afterward" perhaps because "things in the marine realm were a little more insulated," he explains.

Although these estimates might seem rough for such a dramatic event, revealing details on the resolution of years and months "was unimaginable" in decades past, he says. "It's one of the best studied intervals of the geologic record," he notes. And all of this attention has led to increasing nuance in the timeline.

"This is not geologic time—this is instant time," Johnson said, acknowledging that it is a very tricky task to pin down a single event from 65.5 million years ago. But, judging from the chemical, geochemical and geochronological evidence, he said, "The Chicxulub Crater really is the culprit."

5

New Discoveries

Courtesy of the U.S. National Park Service/Photo by Ann Elder
With delicate precision, a researcher hunts for fossils.

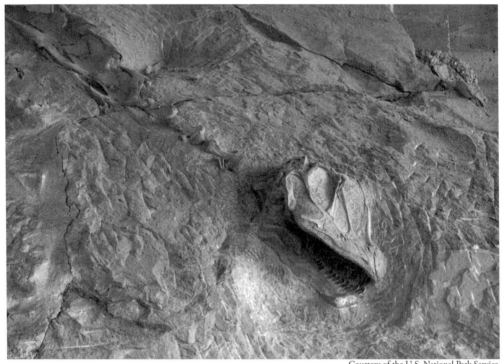

A *Camarasaurus* skull found at Douglass Quarry, part of the Dinosaur National Monument, in eastern Colorado.

Editor's Introduction

Thanks to ongoing fossil finds and technological breakthroughs that allow scientists to examine old evidence in new ways, our picture of how dinosaurs lived is becoming clearer all the time. The articles in this chapter present some of the most exciting discoveries made in recent years.

The chapter begins with "The Hunt for Predator X," James O'Donoghue's look at ancient marine reptiles. Like flying pterosaurs, these creatures were not dinosaurs, but rather a separate branch of the family tree. In addition to reviewing discoveries related to marine reptiles, O'Donoghue shifts the focus to an area of the world not previously addressed in this book: the Arctic.

In "Bringing Up Baby: The Evidence Mounts that Some Dinosaurs Were Attentive Parents," the next piece, David J. Varricchio explains how fossilized egg finds are challenging long-held assumptions regarding how dinosaurs birthed and cared for their offspring. Evidence suggests the omnivorous *Psittacosaurus* and *Maiasauria* fed and looked after their hatchlings, rather than leaving them to fend for themselves. *Oviraptor*, or "egg stealer," meanwhile, has been cleared of the charges that earned it its name. Whereas eggs found alongside adult specimens were once thought to have been swiped from other types of dinosaur, they're now known to be *Oviraptor* eggs—proof of nesting, not theft.

In "Winged Victory," the subsequent article, paleontologist Gareth Dyke presents evidence that neornithines, or modern birds, lived alongside dinosaurs. This hypothesis was first proposed by molecular biologists, who used known rates of genome mutation—"more or less constant," Dyke explains—to count backward and calculate when modern birds likely split from their ancestors on the evolutionary tree. They determined the divergence occurred before the extinction event responsible for killing the dinosaurs—a notion supported by a 70-million-year-old *Teornis* wing found in Mongolia, as well as remnants of a slightly younger genus, *Vegavis*, discovered on Antarctica's Vega Island.

In "Blood from Stone: How Fossils Can Preserve Soft Tissue," Mary H. Schweitzer offers an update on the controversial find—*T. rex* red blood cells—referenced in "Dinosaur Shocker!" a selection from the second chapter. Spurred on by that discovery, Schweitzer has recovered what she claims are bone cells, blood vessels, and other organic materials—remains some peers in the scientific community view with great skepticism. By analyzing soft tissue in the lab, Schweitzer hopes to shed new light on how dinosaurs evolved and ultimately died out.

The book concludes with Richard Stone's "Dinosaurs' Living Descendants," a look at how fossils found in China's Liaoning province are proving the evolutionary link between birds and dinosaurs. With the 2009 discovery of *Anchiornis huxleyi*, scientists were finally able to establish that feathered dinosaurs existed before the earliest known birds. This would need to be the case in order for the latter to have descended from the former.

The Hunt for Predator X[*]

By James O'Donoghue
New Scientist, Oct. 31, 2009

Each summer, a team from the University of Oslo in Norway go hunting for monsters on the island of Spitsbergen. They carry guns in case they get menaced by the world's largest living land carnivore, the polar bear. But it is not bears they are after. They are searching for much bigger quarry, the most formidable predators that ever lived.

Step back 150 million years and Spitsbergen was covered by a cool, shallow sea swarming with marine reptiles. The creatures died out and their fossils became part of an island stuffed full of bones. Nowhere else in the world are so many marine reptiles found in one place.

For a few short weeks the sun never sets and temperatures soar to just above freezing. Knowing that before long the ground will be frozen solid, the researchers dig like crazy. "It's like a gold rush, there are so many fossils waiting to be found," says team leader Jørn Hurum. "The site is densely packed with skeletons. As we speak there are probably more than 1000 skeletons weathering out."

Hurum's Arctic discoveries are part of a remarkable renaissance in interest in the marine reptiles of the Mesozoic era, 251 to 65 million years ago. We now know more about this group of creatures than ever before.

Marine reptiles were among the first vertebrate fossils known to science and were key to the development of the theory of evolution. In the late 18th century the massive jaws of a lizard-like beast were found in a mine in Maastricht in the Netherlands. Later named *Mosasaurus*, the creature helped convince scientists that animals could become extinct, a radical concept in its day. In the early 19th century, ichthyosaurs and plesiosaurs discovered by legendary fossil hunter Mary Anning around Lyme Bay in south-west England helped establish the science of palaeontology. Marine reptiles were among the best-understood extinct creatures

---

MARINE REPTILE BASICS

Marine reptiles are often lumped together with the dinosaurs, but like the flying pterosaurs, they are a separate branch of the family tree . Perhaps the best known are the dolphin-like ichthyosaurs and the plesiosaurs, which looked like the mythical Loch Ness monster. But several other groups evolved over the course of the Mesozoic era (251 to 65 million years ago), most notably pliosaurs and mosasaurs. Ichthyosaurs went extinct about 90 million years ago while the others died out with the dinosaurs.

---

of the first half of the 19th century and played a major role in the intellectual debate nurturing Darwin's theory of evolution.

Yet they faded from view as their terrestrial relatives moved to centre stage. It took nearly a century for marine reptile research to emerge from the shadow cast by the dinosaurs. "Scientists thought they knew all there was to know," says plesiosaur expert Leslie Noè of the Thinktank museum in Birmingham, UK. "The idea was that they weren't worth studying. Nobody would say that now. Our understanding of marine reptiles is phenomenally greater now than it was even 10 years ago."

In the modern world, marine reptiles are few and far between: saltwater crocodiles, turtles and sea snakes are rarities of coastal waters. However, in the ice-free greenhouse of the Mesozoic, reptiles cruised the oceans from pole to pole, occupying the ecological roles now largely filled by whales, dolphins, porpoises, seals and even sharks.

Much like today's marine mammals, marine reptiles evolved from land-living ancestors and were air-breathing. For them, it was a true return to the water. Reptiles evolved around 300 million years ago from amphibian-like ancestors that needed to lay their eggs in water. Reptiles, in contrast, thrive in hot, dry environments.

Among the first to go back were the mesosaurs around 280 million years ago. They were fully aquatic with long, thin bodies, webbed feet and jaws bristling with teeth. They disappeared just a few million years later leaving no known descendants.

Only after the Permian mass extinction 251 million years ago did a full-scale reptilian invasion begin. The extinction was the greatest clear-out of life the world has ever seen and marine life was hit particularly hard: 19 out of every 20 marine species became extinct.

AIR-BREATHERS

The empty seas were ripe for colonisation and reptiles were well placed to take advantage. Temperatures were several degrees warmer than today, which suited cold-blooded reptiles very well. Being air-breathing meant they could thrive in the low-oxygen waters of the post-Permian world where fish struggled to survive. Large predatory fish were also few and far between.

Many types of marine reptile evolved during the Mesozoic, but four stand out owing to their abundance, dominance and global distribution: ichthyosaurs, plesiosaurs, pliosaurs and mosasaurs. All four groups were predatory and included the top marine predators of their time. Some species reached truly enormous sizes.

One key to their success was the evolution of live birth, or vivipary. It has been known for decades that ichthyosaurs reproduced in this way, thanks to well-preserved fossils found at Holzmaden quarry in Germany. One exquisite specimen, now in the State Museum of Natural History in Stuttgart, captures an ichthyosaur in the process of giving birth.

Vivipary was probably seen in all large marine reptiles. In 2001, Mike Caldwell of the University of Alberta, Canada, was examining a mosasaur fossil in the Museum of Natural History in Trieste, Italy. "As soon as I popped open that drawer I knew we had an important discovery. In front of me was a mosasaur with embryos —it had tiny little versions of the adults lined up in its belly," he says.

In 2004 came evidence that a group ancestral to the plesiosaurs, the keichousaurs, also gave birth to live young. As a result, researchers now think that plesiosaurs must also have been viviparous (*Nature*, vol 432, p 383). "Live birth allows you to get much bigger because you don't need to come into shallow water or make your way onto land to lay eggs," says Caldwell. "If you can give birth in water then you can colonise the oceans of the planet" (*Proceedings of the Royal Society of London B*, vol 268, p 2397).

The first big success story was the ichthyosaurs, which appear in the fossil record around 245 million years ago. Early ichthyosaurs were eel-like creatures that stayed close to shore, but over the next 40 million years they evolved into streamlined, dolphin-shaped cruisers that raced through the open oceans, according to ichthyosaur expert Ryosuke Motani of the University of California, Davis (*Nature*, vol 382, p 347).

Some lineages evolved into the biggest marine reptiles that ever lived. In 2004, a team led by Elizabeth Nicholls of the Royal Tyrrell Museum in Drumheller, Canada, excavated a monstrous ichthyosaur from 210-million-year-old rocks in British Columbia. At 21 metres long, *Shonisaurus* was as big as a fin whale, the world's second-largest living animal. "If you blow up a dolphin and make it skinnier then that is probably what *Shonisaurus* looked like," says Motani, who was part of the excavation team (*Journal of Vertebrate Paleontology*, vol 24, p 838). Fragmentary remains suggest that even bigger ichthyosaurs existed around that time.

By the start of the Jurassic period of 200 million years ago, the behemoths were joined by smaller, faster cruisers. In 2002, Motani estimated that *Stenopterygius*, a 180-million-year-old ichthyosaur from Europe, had a cruising speed comparable with tuna, which are among the fastest of all living fish (*Paleobiology*, vol 28, p 251).

The Jurassic was the ichthyosaurs' golden age. They were more abundant than any other marine reptile and were the first group to conquer the deep oceans, as Motani demonstrated through research into the optical properties of their eyes.

In general, eye size and body size are closely correlated in vertebrates: blue whales are the largest living vertebrates and have the biggest eyes, 15 centimetres in diameter.

Many ichthyosaurs bucked that trend. "Ichthyosaur eyes were the biggest of any vertebrate," says Motani. The 4-metre-long *Ophthalmosaurus*, for example, had eyes 23 centimetres across, the size of frisbees, while the eyes of the 9-metre *Temnodontosaurus* were 26 centimetres. Among living creatures, only deep-sea giant squid have eyes of comparable size. Motani argues that giant eyes were an adaptation for diving down 500 metres or more to hunt for squid and other cephalopods, such as the now extinct belemnites.

Motani estimated the visual acuity of ichthyosaur eyes by calculating their light-gathering capacity based on size and focal length. He concluded that they were more sensitive than a typical nocturnal mammal. "At 500 metres down a human would not be able to see a thing but an ichthyosaur would have been able to see moving objects," he says (*Nature*, vol 402, p 747).

At the start of the Jurassic the ichthyosaurs were joined by the plesiosaurs and pliosaurs, which thrived right through until the end of the Cretaceous some 65 million years ago. They were closely related, though they didn't look it: plesiosaurs had long necks, small heads and graceful bodies, while the pliosaurs had massive bodies, short necks and large heads. Both swam using two large pairs of paddles.

Central to the plesiosaurs' biology were their long necks, which in extreme cases could be longer than the rest of the body and tail combined. The neck of *Elasmosaurus* has 72 vertebrae, more than any other animal that we know of. "Long-necked marine animals disappear with the extinction of the plesiosaurs. That way of living just doesn't exist any more," says Noè. Yet long necks were integral to the plesiosaur success story.

Perhaps they were using their long necks to sneak up under schools of fish silhouetted against the sky, suggests marine reptile expert Mike Everhart of the Sternberg Museum of Natural History in Hays, Kansas. "The plesiosaur would have approached from a blind spot as fish can't see well underneath or behind. Then it grabs what it can before the school is alerted." With plesiosaur stomach contents showing fish were a main prey item, this explanation is widely accepted.

However, Noè recently suggested that they were bottom feeders. According to this scenario, the plesiosaur's peg-toothed head rummaged for prey on the sea floor while its body floated above (*Journal of Vertebrate Paleontology*, vol 26, p 105A). Support for this idea came from a 2005 discovery in Queensland, Australia, where Colin McHenry of the University of Newcastle in New South Wales found plesiosaur stomachs full of sea-floor invertebrates (*Science*, vol 310, p 75). McHenry believes that both explanations are correct. "A long neck is a fantastic general-purpose feeding mechanism. It allows you to drift along the bottom and pick out bits that interest you but also gives you the agility to catch fish and squid," he says.

Although plesiosaurs could reach 14 metres, much of their length was taken up by their necks. Overall they were dwarfed by their relatives the pliosaurs, the unquestioned top predators of the Mesozoic seas.

There is some dispute over the identity of the very largest pliosaur, but *Pliosaurus* must come close. It is known from a 3-metre jaw found in Oxfordshire, UK, and Noè estimates that it was up to 18 metres long. "You could put your arm inside its tooth sockets, they are so huge," says Noè, who described the specimen in 2004 (*Proceedings of the Geologists' Association*, vol 115, p 13). He estimates that it weighed as much as 30 tonnes. In comparison, a fully grown *T. rex* was a puny 7 tonnes.

Hurum has found fragments of pliosaurs of similar size in Spitsbergen, two of which—nicknamed "predator X" and "the monster"—could have been as much as 15 metres long.

Not only were they huge, they were also formidable. The stomach contents of an 11-metre Australian pliosaur, *Kronosaurus*, which lived 100 million years ago, reveal it ate plesiosaurs, according to as-yet unpublished research by McHenry. Comparisons with living crocodiles suggest *Kronosaurus* had a much more powerful bite than would be expected for an animal with such a long snout.

For unknown reasons ichthyosaurs and large pliosaurs had died out by 90 million years ago, but it didn't take long for their ecological roles to be refilled.

Mosasaurs were a new breed of marine reptile that branched off from the monitor lizard lineage. Knowledge of the mosasaurs goes back to the discovery of *Mosasaurus*, and their fossil record is more complete than for other marine reptiles. Uniquely, we also know of semi-aquatic transitional forms at the base of the family tree.

Perhaps the best of these "missing links" is the 98-million-year-old *Haasiasaurus*, discovered near Ramallah in the Palestinian West Bank. "*Haasiasaurus* could get around on land just as easily as in the water," says Mike Polcyn of the Southern Methodist University in Dallas, Texas, who described the species in 1999 (*National Science Museum*, Tokyo, Monographs, no 15, p 259).

These early mosasaurs went on to evolve into fully marine forms up to 15 metres long. The final evolutionary radiation of sea monsters had begun and competition was fierce. "Mosasaurs were getting into vicious fights with one another," says Everhart. "I've seen broken bones, crushed skulls and huge bite marks." A 5-metre tylosaur from Kansas that he studied in 2008 was killed by a massive bite to its head. The only animal capable of delivering such an injury was a larger mosasaur, says Everhart (*Transactions of the Kansas Academy of Sciences*, vol 111, p 251).

The very latest mosasaurs showed an interesting evolutionary trend. "Primitive mosasaurs were slender creatures that undulated their bodies like eels," says Johan Lindgren of Lund University in Sweden. "Over time they stiffened their bodies and eventually only swam with their tails, like sharks." This process peaked with *Plotosaurus*, the most advanced mosasaur we know of. In a stunning example of convergent evolution, *Plotosaurus* had evolved a body shape approaching that of the ichthyosaurs (*Lethaia*, vol 40, p 153).

Known only from the latest Cretaceous, the 8-metre-long *Plotosaurus* hints at the way mosasaurs would have evolved—had they not gone extinct.

At the end of the Cretaceous the mosasaurs, plesiosaurs and pliosaurs joined the dinosaurs in the roll call of another mass extinction. "The great marine reptiles were at the top of a long food chain that collapsed 65 million years ago. There was no longer enough food to keep them alive," says Noè.

The sea monsters had had their day. But a vacuum was waiting to be filled, and 10 million years later *Pakicetus*, a carnivorous mammal that looked a bit like a wolf, took a tentative dip in the water. The invasion of the sea had begun again. But that's another story.

Bringing Up Baby[*]

The Evidence Mounts that Some Dinosaurs
Were Attentive Parents

By David J. Varricchio
Natural History, May 2005

In Montana, the summer of 1993 barely existed. It snowed in June and August, and in between there was plenty of cold rain. On ground normally baked hard and dry by the summer sun, I had my first and only badland encounters with salamanders and turtles. For me and my colleagues from Montana State University-Bozeman, the rain also made a mess of our paleontological pursuits, turning the mudstone we were quarrying into its gloppy namesake and preventing glue from holding fossil fragments together. Finally, kept from digging, we prospected for new fossils.

As I picked my way along a relatively firm sandstone ledge, surveying the slippery mudstone, a wet, shiny fossil grabbed my attention. From my fossil-hunting experience in the preceding four summers, I knew it belonged to *Troodon formosus*, a six-foot-long member of the group of dinosaurs, mainly carnivorous, known as theropods. The theropods include such charismatic extinct animals as *Tyrannosaurus rex*, as well as the only surviving dinosaurs—the birds. This *Troodon* was an adult whose bones were still partly articulated, or joined together, a prize compared to the scattered and jumbled remains we were used to finding.

But an even bigger prize was in store. As soon as the rains subsided, we began digging. Beneath the right leg of the animal lay a clutch of at least eight eggs. Was this juxtaposition purely accidental, a result of the vagaries of fossil deposition? Or was it significant? The eggs were of a type ascribed, in the literature of the time, not to *Troodon*, but to a somewhat smaller, herbivorous dinosaur, *Orodromeus*. That identification had been based on the only clutch of such eggs ever found with embryonic remains. (Fossilized embryos are rarely discovered, because their bones only begin to ossify late in development.) Was our *Troodon* caught in the act of raiding an *Orodromeus* nest? Or could the earlier clutch have been misidentified—in which case our find would reflect a parent looking after its own eggs? We could only wonder.

Meanwhile, more than 8,000 miles away in Mongolia, another egg surprise was cooking. Back in the 1920s, explorers from the American Museum of Natural History, in New York City, had discovered clutches of fossil eggs in the Gobi Desert. The explorers had attributed them to the herbivorous dinosaur *Protoceratops*, the most common species in the fossil beds. Their prospecting later turned up a carnivorous theropod dinosaur on top of a clutch of the same eggs, and—drawing what seemed to be the obvious conclusion—they had named it *Oviraptor*, "egg stealer." Now, in 1993, on a new American Museum expedition, paleontologist Mark A. Norell was discovering that both of the previous conclusions were wrong.

As any aficionado of dinosaurs has heard by now, what Norell and company found was a fossil embryo preserved in a supposed *Protoceratops* egg—except that the embryo was an *Oviraptor*. The original identification of the eggs was incorrect. Moreover, poor *Oviraptor*, maligned as an egg stealer, was apparently brooding its own nest as subsequent discoveries by American Museum and Sino-Canadian expeditions have confirmed. Clearly these dinosaurs, at least, cared for their eggs.

The new find in the Gobi, and the subsequent rehabilitation of *Oviraptor*'s image, encouraged me to take a closer look at our own *Troodon* fossil for clues of parental care. As I've pursued that whodunit, new material that may demonstrate parental care in yet other fossil dinosaurs has come to light. Paleontologists have long attributed the evolutionary successes of mammals to a variety of features, but a critically important one has been parental care. After all, the name of the group comes from the mammary glands, the organ virtually synonymous with parental care. But the new dinosaur discoveries suggest that perhaps it's time we stopped looking down our mammalian noses at the creatures that didn't survive the Cretaceous mass extinction.

As early as 1979, John R. ("Jack") Horner, also of Montana State University-Bozeman, proposed the radical idea that some dinosaurs not only attended their eggs but cared for their young as well. In western Montana, along the Rocky Mountain Front, Horner's team had uncovered grapefruit-size eggs and two groups of young of a new herbivorous species of hadrosaur, or duck-billed dinosaur, which they named *Maiasaura*. The fossilized young were contained in bowl-shaped sedimentary structures, which he interpreted as nests. One of the groups was intermixed with the fossils of broken eggshells. Judging by the size of the animals—only eighteen inches long—compared with the eggs, these *Maiasaura* offspring had been either embryos close to hatching or newly hatched young. The young in the second group were substantially older individuals, each about three feet long.

Horner concluded that these *Maiasaura* young had remained nest-bound and dependent on adults for food. His argument was supported by what he found when he looked at the fossils under the microscope. The ends of the limb bones had been cartilaginous at the time of death. Such immature limbs would have been too weak for the young animals to have run about on their own. Similar growth patterns occur in the nest-bound young of some birds.

Horner argued that dinosaurs were not the cold-blooded (in the emotional as well as thermal sense), uncaring parents they had previously been assumed to be. But this break with orthodoxy meant that his evidence would be disputed. At the time, paleontologists were stuck in a reptilian perspective on dinosaurs. Few had considered that dinosaur reproduction might better reflect that of crocodilians and birds, the extinct dinosaurs' closest living relatives.

Both crocodilians and birds exhibit fairly extensive parental care. Crocodilians guard their nests, help their young to hatch, and even protect them during early development. Many familiar birds, such as the backyard robin, feed their young. There is also a wide range of birds, from chickens to ostriches, whose young are able to feed on their own as soon as they are hatched, but which still depend on their parents for protection.

By the early 1990s, Horner's ideas had gained a more receptive audience. When the news broke about *Oviraptor* and its misidentified eggs, our suspicions about our own fossil grew. Step one was to reevaluate the identification of the eggs. Fortunately, during the previous summers we had amassed the largest known sample of *Troodon* remains anywhere, from a rich bone bed discovered by Horner [*see "The Birthday Site," by David J. Varricchio, April 1997*]. Armed with this new material, we examined the original embryos.

The *Orodromeus* identification had indeed been mistaken: our eggs belonged to *Troodon*. In hindsight, several factors had contributed to the earlier error. One was the prevalence of *Orodromeus* fossils near the eggs that contained the embryos (a false lead, as in the case of *Protoceratops* and *Oviraptor*). Another factor was the earlier paucity of good *Troodon* material for comparison with the embryos. And finally, because the embryos had poorly developed teeth, they lacked the most characteristic *Troodon* feature—teeth with unusually large and distinctive serrations.

Another piece of the *Troodon* puzzle fell into place the following summer. In western Montana, about seventy miles south of our 1993 *Troodon* find, we came upon a complete clutch of twenty-four *Troodon* eggs. The eggs rested nearly vertically in the ground, the upper portions within a soft mudstone and the bottoms within soil that over the ages had hardened into a limestone. Our awls and ice picks easily stripped away the mudstone but were completely ineffectual against the limestone. After we worked around the eggs, we were left with a shallow bowl of limestone with a raised rim—the remains of a *Troodon* nest built some 75 million years ago. The structure suggests an open nest with the upper parts of the eggs exposed, and brooding by an attending adult.

A couple of complete clutches and a dozen partial ones are now known for *Troodon*. The half-buried eggs, and those of *Oviraptor* as well (also now known to be half buried), are good evidence that the egg-laying behavior of these dinosaurs was innovative. To this day, many reptiles bury their eggs, whereas nearly all birds leave them exposed. The theropods seem to have been pointing the way toward the present-day behavior of birds. *Troodon*'s eggs also had a large end and a small end, as do the eggs of modern birds. Research I undertook with Frankie D. Jackson of Montana State University-Bozeman, who specializes in the study of eggs and nest sites, shows that these extinct theropod egg shells were nearly indistinguishable at the microscopic level from those of modern birds.

Curiously, the eggs in the clutches occur in pairs, as do eggs in some other theropod clutches, suggesting the female laid two eggs at each sitting. Modern birds also lay intermittently, but because they only have one functional ovary and oviduct, they lay just one egg at a time. In contrast, modern turtles and crocodilians lay all their eggs in one bout. The paired eggs suggest that these dinosaurs retained two functioning ovaries and oviducts, each producing a single egg at a time.

Yet there is no hard evidence that the theropods cared for their hatchlings. The bones of their embryos, in contrast to those of Horner's *Maiasaura* hatchlings, appear well ossified, indicating that the theropod young came into the world ready to forage for themselves.

A new case for parental care of young dinosaurs, including their feeding, comes from the paleontologist Robert T. Bakker, who is legendary for his provocative ideas. Bakker recently examined some sites in Wyoming that include isolated bones of large herbivorous dinosaurs, as well as the shed teeth of both small and large individuals of the carnivorous theropod *Allosaurus*. His conclusion: The sites were the lairs of *Allosaurus*, places to which adults brought food to feed their offspring. I find his arguments intriguing but not convincing. First, these "lairs" occur in what would have been exposed floodplains. Second, Bakker maintains that the bones and teeth were in place before they were covered with the sediments that now entomb them. A more standard explanation, though, seems sufficient: water simply carried and deposited the fossils and sediments together.

A much more convincing find, from Liaoning Province in northeastern China, presses the point of parental care much more effectively. In 2003 I had the good fortune to be asked by Timothy Huang of Paleoworld-Taiwan and Liu Jinyuan and Gao Chunling of the Dalian Natural History Museum, in Dalian, China, to take part in describing an exceptional specimen.

The specimen includes the partial remains of a single adult *Psittacosaurus* surrounded by thirty-four young, each about nine inches long and most of them complete, their skeletons articulated and essentially upright. The animals crowd into a shallow bowl-shaped depression less than a yard across. The excellent preservation of the skeletons indicates that the animals were buried rapidly and rules out the possibility that they were somehow swept together by the elements. But could such a large group have been a mother and her brood?

No one yet knows enough about *Psittacosaurus* reproduction, such as egg, clutch, and hatchling size, to say for sure. If this dinosaur had small eggs, a brood of thirty-four may have been a reasonable possibility. Yet several other circumstances might account for the apparently large number of young. Perhaps one male protected the eggs laid by his multiple female partners. Or perhaps the young from various females were typically gathered after hatching to form crèches, requiring fewer adults to watch over them.

Psittacosaurus and Horner's *Maiasaura* provide the best evidence to date for the parental care of hatchlings and young among extinct dinosaurs. Another good candidate is *Protoceratops*, for which groups of small young have been found, but so far without an attending adult. All three are herbivorous dinosaurs. Although some extinct carnivorous theropods brooded their eggs, as do their relatives the birds, there is sparse evidence that such theropods minded or fed their young. It's conceivable that the contrast reflects a difference in the creatures' ecological niches. But with so little hard evidence, it's better not to jump to any conclusions.

Paleontologists have learned that they must take care to understand extinct species on their own terms. What no one disputes, though, is that reproductive strategies are directly linked to evolutionary success. Parental care in its various forms may have been among the key adaptations that enabled these remarkable species, now reduced to fossils, to have prevailed for so long.

Winged Victory[*]

Winged Birds Now Found to Have Been Contemporaries of Dinosaurs

By Gareth Dyke
Scientific American, July 2010

December in Moscow, and the temperature drops under 15 degrees below zero. The radiators in the bar have grown cold, so I sit in a thick coat and gloves drinking vodka while I ponder the fossil birds. The year is 2001, and Evgeny N. Kurochkin of the Russian Academy of Sciences and I have just spent hours at the paleontology museum as part of our effort to survey all the avian fossils ever collected in Mongolia by joint Soviet-Mongolian expeditions. Among the remains is a wing unearthed in the Gobi Desert in 1987. Compared with the spectacularly preserved dinosaur skeletons in the museum's collections, this tiny wing—its delicate bones jumbled and crushed—is decidedly unglamorous. But it offers a strong hint that a widely held view of bird evolution is wrong.

More than 10,000 species of birds populate the earth today. Some are adapted to living far out on the open ocean, others eke out a living in arid deserts, and still others dwell atop snow-capped mountains. Indeed, of all the classes of land vertebrates, the one comprising birds is easily the most diverse. Evolutionary biologists long assumed that the ancestors of today's birds owed their success to the asteroid impact that wiped out the dinosaurs and many other land vertebrates around 65 million years ago. Their reasoning was simple: although birds had evolved before that catastrophe, anatomically modern varieties appeared in the fossil record only after that event. The dawning of ducks, cuckoos, hummingbirds and other modern forms—which together make up the neornithine ("new birds") lineage—seemed to be a classic case of an evolutionary radiation in response to the clearing out of ecological niches by an extinction event. In this case, the niches were those occupied by dinosaurs, the flying reptiles known as pterosaurs and archaic birds.

KEY CONCEPTS

- The descent of birds from small, meat-eating dinosaurs is by now established. Far less clear is the origin of anatomically modern birds.
- The conventional fossil-based thinking is that modern birds arose only after the asteroid impact that claimed the dinosaurs and many other creatures 65 million years ago.
- But molecular studies and a smattering of equivocal fossil finds have hinted that modern birds might have deeper roots.
- Recently analyzed fossils of ancient modern birds confirm this earlier origin, raising the question of why these birds, but not the archaic ones, survived the mass extinction.

Over the past decade, however, mounting evidence from the fossil record—including that crushed wing—and from analyses of the DNA of living birds has revealed that neornithine birds probably diversified earlier than 65 million years ago. The findings have upended the traditional view of bird evolution—and sparked important new questions about how these animals soared to evolutionary heights.

EARLY BIRDS

Birds are one of just three groups of vertebrates ever to have evolved active, flapping flight. The other two are the ill-fated pterosaurs and the bats, which appeared much later and share the skies with birds to this day. For years paleontologists debated the origin of the earliest birds. One side argued that they evolved from small, meat-eating dinosaurs called theropods; the other contended that they evolved from earlier reptiles. But the discoveries over the past two decades of birdlike dinosaurs, including many with downy coats, have convinced most scientists that birds evolved from theropod dinosaurs.

Connecting the dots between ancestral avians and modern birds has proved far trickier, however. Consider *Archaeopteryx*, the 145-million-year-old creature from Germany that is the oldest known bird. *Archaeopteryx* preserves the earliest definitive evidence for wings with asymmetric feathers capable of generating the lift required for flight—one defining characteristic of the group. Yet it more closely resembles small-bodied dinosaurs such as *Velociraptor*, *Deinonychus*, *Anchiornis* and *Troodon* than modern birds. Like those dinosaurs, early birds such as *Archaeopteryx* and the more recently discovered *Jeholornis* from China and *Rahonavis* from Madagascar possessed long, bony tails, and some had sharp teeth, among other primitive traits. Neornithines, in contrast, lack those characteristics and exhibit a suite of advanced ones. These features include fully fused toe bones and fingerless wings, which reduce the weight of the skeleton, allowing more efficient flight, and highly flexible wrists and wings, which enhance maneuverability in the air. How and when the neornithines acquired these traits were impossible to determine, however, thanks to an absence of fossils documenting the transition.

This is not to say the fossil record lacked avian remains intermediate in age between the first birds and the postextinction neornithines. Clearly by the early Cretaceous, more than 100 million years ago, birds representing a wide range of flight adaptations and ecological specializations had evolved. Some flew on wings that were broad and wide; others had wings that were long and thin. Some lived in forests eating insects and fruit; others made their home along lakeshores or in the water and subsisted on fish. This incredible diversity persisted through the latest stages of the Cretaceous, 65 million years ago. In fact, along with my Dutch colleagues at the Natural History Museum in Maastricht, I have described remains of toothed birds found just below the geologic horizon that marks the end-Cretaceous extinction event. But all the Cretaceous birds complete enough to classify belonged to lineages more ancient than neornithines, and these lineages did not survive the catastrophe—which is why, until recently, the available evidence implied that the simplest explanation for the rise of modern birds was that they originated and radiated after the extinction event.

MOLECULAR CLUES

By the 1990s, while paleontologists were still looking for ancestral neornithines in the Cretaceous and coming up empty-handed, another method of reconstructing the evolutionary history of organisms—one that did not involve the fossil record—was gaining traction. Molecular biologists were sequencing the DNA of living organisms and comparing those sequences to estimate when two groups split from each other. They can make such estimates because certain parts of the genome mutate at a more or less constant rate, constituting the "ticking" of the so-called molecular clock.

Molecular biologists had long questioned the classical, fossil-based view of modern bird evolution. So they tackled the problem using their clock technique to estimate the divergence dates for major lineages of modern birds. Among the most significant splits is the one that occurred between the large, mostly flightless paleognaths (ostriches and emu and their kin) and the Galloanserae (which includes chickens and other members of the Galliformes group, as well as ducks and other members of the Anseriformes group). The DNA studies concluded that these two lineages—the most primitive of the living neornithines—split from each other deep in the Cretaceous. And researchers obtained similarly ancient divergence dates for other lineages.

The findings implied that, contrary to conventional paleontological wisdom, neornithines lived alongside dinosaurs. It is funny to think of a robin perched on the back of a *Velociraptor* or a duck paddling alongside a *Spinosaurus*. But the molecular evidence for the contemporaneity of modern birds and dinosaurs was so compelling that even the paleontologists—who have typically viewed with skepticism those DNA findings that conflict with the fossil record—began to embrace it.

Still, those of us who study ancient skeletons urgently wanted fossil confirmation of this new view of bird evolution.

DUCKS IN A ROW

At last, after the new millennium, paleontologists' luck began to change for the better, starting with the tiny Mongolian wing that Evgeny and I focused on in Moscow. Back when Evgeny first saw the fossil in 1987, he told me that he thought it looked like a member of the presbyornithids, a group of now extinct ducklike birds related to modern ducks and geese. But at 70 million years old, it was a Cretaceous bird, and everyone knew—or thought they did—that there was no definitive evidence for presbyornithids in the Cretaceous. Yet our comparisons in the museum that cold winter in 2001 demonstrated conclusively that the wing—with its straight carpometacarpus (the bone formed by the fusion of the hand bones) and details of canals, ridges and muscle scars—did indeed belong to a presbyornithid, which, moreover, was the oldest unequivocal representative of any neornithine group. Our finding fit the predictions of the molecular biologists perfectly. In a 2002 paper that formally described the animal, we gave it the name *Teviornis*.

Before long, *Teviornis* was joined by a second confirmed early neornithine, *Vegavis*, from Antarctica's Vega Island. *Vegavis* had been found in the 1990s only to languish in relative anonymity for years before its true significance came to light. In 2005 Julia A. Clarke, now at the University of Texas at Austin, and her colleagues published a paper showing that *Vegavis* was another bird from the Cretaceous that exhibits a number of features found in modern ducks, particularly in its broad shoulder girdle, pelvis, wing bones and lower legs. At 66 million to 68 million years old, *Vegavis* is a little younger than *Teviornis* but still clearly predates the mass extinction. And it is a much more complete fossil, preserving the better part of a skeleton.

For most paleontologists, *Vegavis* clinched the case for Cretaceous neornithines. Thus enlightened, researchers have begun reexamining fossil collections from this time period, looking for additional examples of early modern birds. One investigator, Sylvia Hope of the California Academy of Sciences in San Francisco, had been arguing for years that bird species she has identified from fossils found in New Jersey and Wyoming that date to between 80 million and 100 million years ago are modern. But the finds—mostly single bones—had been considered by other researchers as too scrappy to identify conclusively. The revelations about *Vegavis* and *Teviornis* suggest that she was right all along. Comparisons of Hope's bones with more complete remains should prove illuminating in this regard.

FLYING THE COOP

Rooting modern birds in the Cretaceous neatly aligned the fossil record with the DNA-based divergence dates. But it raised a vexing new question, namely, Why were modern birds able to survive the asteroid impact and its attendant ecological changes when their more primitive avian cousins and their fellow fliers, the pterosaurs, were not? To my mind, this constitutes the single biggest remaining mystery of bird evolution. The answer is still very much up for grabs, and I am devoting much of my research at the moment to trying to get at it.

With only a couple of confirmed Cretaceous neornithines on record, there is not much in the way of fossil clues to go on. Insights have come from studies of living birds, however. Using a huge data set of measurements of living birds, my colleagues in the U.K. and I have shown, for example, that the wing-bone proportions of primitive modern birds, including *Teviornis* and *Vegavis*, are no different from those of the extinct enantiornithines. Comparing the fossil wing-bone proportions with those of living birds allows us to infer some aspects of wing shape and hence gain information about the aerodynamic capabilities of fossil birds. But so far as we can tell, the wing shapes of the two groups of fossil birds do not differ; in other words, we do not think that early neornithines were any better at flying than were the enantiornithines (although both these groups were most likely better in the air than earlier theropodlike birds such as *Archaeopteryx*).

If flight ability did not give the neornithines an advantage over their Cretaceous counterparts, what did? A number of paleontologists, including me, have posited that differences in foraging habits might have conferred a competitive edge. In support of that theory, I have shown in a series of papers published over the past few years that modern birds preserved in the immediate aftermath of the mass extinction, in rocks 60 million years old and younger, probably lived mostly in wet environments: coastlines, lakes, the edges of rivers and the deep ocean, for example. Many of the birds that inhabit such environments today—ducks among them— are typically generalists, able to subsist on a wide variety of foods. And ducklike birds are currently the one confirmed lineage of modern birds we have found in the Cretaceous. The groups of Cretaceous birds that did not survive the disaster, in contrast, have been collected from rocks that were formed in many different kinds of environments—including seashores, inland areas, deserts and forests. This ecological diversity may indicate that the archaic birds had evolved specializations for feeding in each of these niches. Perhaps, then, the secret of early modern birds' success was simply the fact that they were less specialized than the other groups.

Such flexibility might have enabled the neornithines to adapt more easily to the changing conditions that followed the asteroid impact. It is an appealing idea, but these are early days. Only with the discovery of more fossils—whether in the ground or in museum drawers—will we be able to determine how modern birds eluded elimination and took wing.

GARETH DYKE *prefers dry bones and flattened fossils to living birds. A paleontologist at University College Dublin, he became interested in animal flight when he was an undergraduate student in England. In conducting his research on the evolution of birds and their flight, he has studied and described fossils from all over the world. When not traveling to visit museums or to do fieldwork in the middle of deserts, Dyke enjoys learning about 19th-century European history. He is writing a book about a Transylvanian* dinosaur *collector who was also a spy for Austria-Hungary.*

MORE TO EXPLORE

A New Presbyornithid Bird (Aves, Anseriformes) from the Late Cretaceous of Southern Mongolia. E. N. Kurochkin, G.J. Dyke and A. A. Karhu in *American Museum Novitates*, No. 3866, pages 1–12; December 27, 2002.

Survival in the First Hours of the Cenozoic. Douglas S. Robertson et al. in *Geological Society of America Bulletin*, Vol. 116, Nos. 5–6, pages 760–768; May 2004.

Definitive Fossil Evidence for the Extant Avian Radiation in the Cretaceous. Julia A. Clarke et al. in *Nature*, Vol. 433, pages 305–308; January 20, 2005.

The Beginnings of Birds: Recent Discoveries, Ongoing Arguments and New Directions. Luis M. Chiappe and Gareth J. Dyke in *Major Transitions in Vertebrate Evolution*. Edited by J. S. Anderson and H.D. Sues. Indiana University Press, 2007.

The Inner Bird: Anatomy and Evolution. Gary W. Kaiser. University of British Columbia Press, 2007.

Blood from Stone*

How Fossils Can Preserve Soft Tissue

By Mary H. Schweitzer
Scientific American, December 2010

Peering through the microscope at the thin slice of fossilized bone, I stared in disbelief at the small red spheres a colleague had just pointed out to me. The tiny structures lay in a blood vessel channel that wound through the pale yellow hard tissue. Each had a dark center resembling a cell nucleus. In fact, the spheres looked just like the blood cells in reptiles, birds and all other vertebrates alive today except mammals, whose circulating blood cells lack a nucleus. They couldn't be cells, I told myself. The bone slice was from a dinosaur that a team from the Museum of the Rockies in Bozeman, Mont., had recently uncovered—a *Tyrannosaurus rex* that died some 67 million years ago—and everyone knew organic material was far too delicate to persist for such a vast stretch of time.

For more than 300 years paleontologists have operated under the assumption that the information contained in fossilized bones lies strictly in the size and shape of the bones themselves. The conventional wisdom holds that when an animal dies under conditions suitable for fossilization, inert minerals from the surrounding environment eventually replace all of the organic molecules—such as those that make up cells, tissues, pigments and proteins—leaving behind bones composed entirely of mineral. As I sat in the museum that afternoon in 1992, staring at the crimson structures in the dinosaur bone, I was actually looking at a sign that this bedrock tenet of paleontology might not always be true—though at the time, I was mostly puzzled. Given that dinosaurs were nonmammalian vertebrates, they would have had nucleated blood cells, and the red items certainly looked the part, but so, too, they could have arisen from some geologic process unfamiliar to me.

Back then, I was a relatively new graduate student at Montana State University, studying the microstructure of dinosaur bone, hardly a seasoned pro. After I sought opinions on the identity of the red spheres from faculty members and other graduate students, word of the puzzle reached Jack Horner, curator of paleontology at the museum and one of the world's foremost dinosaur authorities. He took a look for himself. Brows furrowed, he gazed through the microscope for what seemed like hours without saying a word. Then, looking up at me with a frown, he asked, "What do *you* think they are?" I replied that I did not know, but they were the right size, shape and color to be blood cells, and they were in the right place, too. He grunted. "So prove to me they aren't." It was an irresistible challenge, and one that has helped frame how I ask my research questions, even now.

Since then, my colleagues and I have recovered various types of organic remains—including blood vessels, bone cells and bits of the fingernail-like material that makes up claws—from multiple specimens, indicating that although soft-tissue preservation in fossils may not be common, neither is it a one-time occurrence. These findings not only diverge from textbook description of the fossilization process, they are also yielding fresh insights into the biology of bygone creatures. For instance, bone from another *T. rex* specimen has revealed that the animal was a female that was "in lay" (preparing to lay eggs) when she died—information we could not have gleaned from the shape and size of the bones alone. And a protein detected in remnants of fibers near a small carnivorous dinosaur unearthed in Mongolia has helped establish that the dinosaur had feathers that, at the molecular level, resembled those of birds.

Our results have met with a lot of skepticism—they are, after all, extremely surprising. But the skepticism is a proper part of science, and I continue to find the work fascinating and full of promise. The study of ancient organic molecules from dinosaurs has the potential to advance understanding of the evolution and extinction of these magnificent creatures in ways we could not have imagined just two decades ago.

FIRST SIGNS

Extraordinary claims, as the old adage goes, require extraordinary evidence. Careful scientists make every effort to disprove cherished hypotheses before they accept that their ideas are correct. Thus, for the past 20 years I have been trying every experiment I can think of to disprove the hypothesis that the materials my collaborators and I have discovered are components of soft tissues from dinosaurs and other long-gone animals.

In the case of the red microstructures I saw in the *T. rex* bone, I started by thinking that if they were related to blood cells or to blood cell constituents (such as molecules of hemoglobin or heme that had clumped together after being released from dying blood cells), they would have persisted in some, albeit possibly very altered, form only if the bones themselves were exceptionally well preserved. Such tissue would have disappeared in poorly preserved skeletons. At the macroscopic level, this was clearly true. The skeleton, a nearly complete specimen from eastern Montana—officially named MOR 555 and affectionately dubbed "Big Mike"—

includes many rarely preserved bones. Microscope examination of thin sections of the limb bones revealed similarly pristine preservation. Most of the blood vessel channels in the dense bone were empty, not filled with mineral deposits as is usually the case with dinosaurs. And those ruby microscopic structures appeared only in the vessel channels, never in the surrounding bone or in sediments adjacent to the bones, just as should be true of blood cells.

Next, I turned my attention to the chemical composition of the blood cell look-alikes. Analyses showed that they were rich in iron, as red blood cells are, and that the iron was specific to them. Not only did the elemental makeup of the mysterious red things (we nicknamed them LRRTs, "little round red things") differ from that of the bone immediately surrounding the vessel channels, it was also utterly distinct from that of the sediments in which the dinosaur was buried. But to further test the connection between the red structures and blood cells, I wanted to examine my samples for heme, the small iron-containing molecule that gives vertebrate blood its scarlet hue and enables hemoglobin proteins to carry oxygen from the lungs to the rest of the body. Heme vibrates, or resonates, in telltale patterns when it is stimulated by tuned lasers, and because it contains a metal center, it absorbs light in a very distinct way. When we subjected bone samples to spectroscopy tests—which measure the light that a given material emits, absorbs or scatters—our results showed that somewhere in the dinosaur's bone were compounds that were consistent with heme.

One of the most compelling experiments we conducted took advantage of the immune response. When the body detects an invasion by foreign, potentially harmful substances, it produces defensive proteins called antibodies that can specifically recognize, or bind to, those substances. We injected extracts of the dinosaur bone into mice, causing the mice to make antibodies against the organic compounds in the extract. When we then exposed these antibodies to hemoglobin from turkeys and rats, they bound to the hemoglobin—a sign that the extracts that elicited antibody production in the mice had included hemoglobin or something very like it. The antibody data supported the idea that Big Mike's bones contained something similar to the hemoglobin in living animals.

None of the many chemical and immunological tests we performed disproved our hypothesis that the mysterious red structures visible under the microscope were red blood cells from a *T. rex*. Yet we could not show that the hemoglobin-like substance was specific to the red structures—the available techniques were not sufficiently sensitive to permit such differentiation. Thus, we could not claim definitively that they were blood cells. When we published our findings in 1997, we drew our conclusions conservatively, stating that hemoglobin proteins might be preserved and that the most likely source of such proteins was the cells of the dinosaur. The paper got very little notice.

THE EVIDENCE BUILDS

Through the *T. rex* work, I began to realize just how much fossil organics stood to reveal about extinct animals. If we could obtain proteins, we could conceivably decipher the sequence of their constituent amino acids, much as geneticists sequence the "letters" that make up DNA. And like DNA sequences, protein sequences contain information about evolutionary relationships between animals, how species change over time and how the acquisition of new genetic traits might have conferred advantages to the animals possessing those features. But first I had to show that ancient proteins were present in fossils other than the wonderful *T. rex* we had been studying. Working with Mark Marshall, then at Indiana University, and with Seth Pincus and John Watt, both at Montana State during this time, I turned my attention to two well-preserved fossils that looked promising for recovering organics.

The first was a beautiful primitive bird named *Rahonavis* that paleontologists from Stony Brook University and Macalester College had unearthed from deposits in Madagascar dating to the Late Cretaceous period, around 80 million to 70 million years ago. During excavation they had noticed a white, fibrous material on the skeleton's toe bones. No other bone in the quarry seemed to have the substance, nor was it present on any of the sediments there, suggesting that it was part of the animal rather than having been deposited on the bones secondarily. They wondered whether the material might be akin to the strong sheath made of keratin protein that covers the toe bones of living birds, forming their claws, and asked for my assistance.

Keratin proteins are good candidates for preservation because they are abundant in vertebrates, and the composition of this protein family makes them very resistant to degradation—something that is nice to have in organs such as skin that are exposed to harsh conditions. They come in two main types: alpha and beta. All vertebrates have alpha keratin, which in humans makes up hair and nails and helps the skin to resist abrasion and dehydration. Beta keratin is absent from mammals and occurs only in birds and reptiles among living organisms.

To test for keratins in the white material on the *Rahonavis* toe bones, we employed many of the same techniques I had used to study *T. rex*. Notably, antibody tests indicated the presence of both alpha and beta keratin. We also applied additional diagnostic tools. Other analyses, for instance, detected amino acids that were localized to the toe-bone covering and also detected nitrogen (a component of amino acids) that was bound to other compounds much as proteins bind together in living tissues, including in keratin. The results of all our tests supported the notion that the cryptic white material covering the ancient bird's toe bones included fragments of alpha and beta keratin and was the remainder of its once lethal claws.

The second specimen we probed was a spectacular Late Cretaceous fossil that researchers from the American Museum of Natural History in New York City had discovered in Mongolia. Although the scientists dubbed the animal *Shuvuuia deserti*, or "desert bird," it was actually a small carnivorous dinosaur. While cleaning the fossil, Amy Davidson, a technician at the museum, noticed small white fibers in the animal's neck region. She asked me if I could tell if they were remnants of feathers. Birds are descended from dinosaurs, and fossil hunters have discovered a number of dinosaur fossils that preserve impressions of feathers, so in theory the suggestion that *Shuvuuia* had a downy coat was plausible. I did not expect that a structure as delicate as a feather could have endured the ravages of time, however. I suspected the white fibers instead came from modern plants or from fungi. But I agreed to take a closer look.

To my surprise, initial tests ruled out plants or fungi as the source of the fibers. Moreover, subsequent analyses of the microstructure of the strange white strands pointed to the presence of keratin. Mature feathers in living birds consist almost exclusively of beta keratin. If the small fibers on *Shuvuuia* were related to feathers, then they should harbor beta keratin alone, in contrast to the claw sheath of *Rahonavis*, which contained both alpha and beta keratin. That, in fact, is exactly what we found when we conducted our antibody tests—results we published in 1999.

EXTRAORDINARY FINDS

By now I was convinced that small remnants of original proteins could survive in extremely well preserved fossils and that we had the tools to identify them. But many in the scientific community remained unconvinced. Our findings challenged everything scientists thought they knew about the breakdown of cells and molecules. Test-tube studies of organic molecules indicated that proteins should not persist more than a million years or so; DNA had an even shorter life span. Researchers working on ancient DNA had claimed previously that they had recovered DNA millions of years old, but subsequent work failed to validate the results. The only widely accepted claims of ancient molecules were no more than several tens of thousands of years old. In fact, one anonymous reviewer of a paper I had submitted for publication in a scientific journal told me that this type of preservation was not possible and that I could not convince him or her otherwise, regardless of our data.

In response to this resistance, a colleague advised me to step back a bit and demonstrate the efficacy of our methods for identifying ancient proteins in bones that were old, but not as old as dinosaur bone, to provide a proof of principle. Working with analytical chemist John Asara of Harvard University, I obtained proteins from mammoth fossils that were estimated to be 300,000 to 600,000 years old. Sequencing of the proteins using a technique called mass spectrometry identified them unambiguously as collagen, a key component of bone, tendons, skin and other tissues. The publication of our mammoth results in 2002 did not trigger much controversy. Indeed, the scientific community largely ignored it. But our proof of principle was about to come in very handy.

The next year a crew from the Museum of the Rockies finally finished excavating another *T. rex* skeleton, which at 68 million years old is the oldest one to date. Like the younger *T. rex*, this one—called MOR 1125 and nicknamed "Brex," after discoverer Bob Harmon—was recovered from the Hell Creek Formation in eastern Montana. The site is isolated and remote, with no access for vehicles, so a helicopter ferried plaster jackets containing excavated bones from the site to the camp. The jacket containing the leg bones was too heavy for the helicopter to lift. To retrieve them, then, the team broke the jacket, separated the bones and rejacketed them. But the bones are very fragile, and when the original jacket was opened, many fragments of bone fell out. These were boxed up for me. Because my original *T. rex* studies were controversial, I was eager to repeat the work on a second *T. rex*. The new find presented the perfect opportunity.

As soon as I laid eyes on the first piece of bone I removed from that box, a fragment of thighbone, I knew the skeleton was special. Lining the internal surface of this fragment was a thin, distinct layer of a type of bone that had never been found in dinosaurs. This layer was very fibrous, filled with blood vessel channels, and completely different in color and texture from the cortical bone that constitutes most of the skeleton. "Oh, my gosh, it's a girl—and it's pregnant!" I exclaimed to my assistant, Jennifer Wittmeyer. She looked at me like I had lost my mind. But having studied bird physiology, I was nearly sure that this distinctive feature was medullary bone, a special tissue that appears for only a limited time (often for just about two weeks), when birds are in lay, and that exists to provide an easy source of calcium to fortify the eggshells.

One of the characteristics that sets medullary bone apart from other bone types is the random orientation of its collagen fibers, a characteristic that indicates very rapid formation. (This same organization occurs in the first bone laid down when you have a fracture—that is why you feel a lump in healing bone.) The bones of a modern-day bird and all other animals can be de-mineralized using mild acids to reveal the telltale arrangement of the collagen fibers. Wittmeyer and I decided to try to remove the minerals. If this was medullary bone and if collagen was present, eliminating the minerals should leave behind randomly oriented fibers. As the minerals were removed, they left a flexible and fibrous clump of tissue. I could not believe what we were seeing. I asked Wittmeyer to repeat the experiment multiple times. And each time we placed the distinctive layer of bone in the mild acid solution, fibrous stretchy material remained—just as it does when medullary bone in birds is treated in the same way.

Furthermore, when we then dissolved pieces of the denser, more common cortical bone, we obtained more soft tissue. Hollow, transparent, flexible, branching tubes emerged from the dissolving matrix—and they looked exactly like blood vessels. Suspended inside the vessels were either small, round red structures or amorphous accumulations of red material. Additional demineralization experiments revealed distinctive-looking bone cells called osteocytes that secrete the collagen and other components that make up the organic part of bone. The whole dinosaur seemed to preserve material never seen before in dinosaur bone.

When we published our observations in *Science* in 2005, reporting the presence of what looked to be collagen, blood vessels and bone cells, the paper garnered a lot of attention, but the scientific community adopted a wait-and-see attitude. We claimed only that the material we found resembled these modern components—not that they were one and the same. After millions of years, buried in sediments and exposed to geochemical conditions that varied over time, what was preserved in these bones might bear little chemical resemblance to what was there when the dinosaur was alive. The real value of these materials could be determined only if their composition could be discerned. Our work had just begun.

Using all the techniques honed while studying Big Mike, *Rahonavis*, *Shuvuuia* and the mammoth, I began an in-depth analysis of this *T. rex*'s bone in collaboration with Asara, who had refined the purification and sequencing methods we used in the mammoth study and was ready to try sequencing the dinosaur's much older proteins. This was a much harder exercise, because the concentration of organics in the dinosaur was orders of magnitude less than in the much younger mammoth and because the proteins were very degraded. Nevertheless, we were eventually able to sequence them. And, gratifyingly, when our colleague Chris Organ of Harvard compared the *T. rex* sequences with those of a multitude of other organisms, he found that they grouped most closely with birds, followed by crocodiles—the two groups that are the closest living relatives of dinosaurs.

CONTROVERSY AND ITS AFTERMATH

Our papers detailing the sequencing work, published in 2007 and 2008, generated a firestorm of controversy, most of which focused on our interpretations of the sequencing (mass spectrometry) data. Some dissenters charged that we had not produced enough sequences to make our case; others argued that the structures we interpreted as primeval soft tissues were actually biofilm—"slime" produced by microbes that had invaded the fossilized bone. There were other criticisms, too. I had mixed feelings about their feedback. On one hand, scientists are paid to be skeptical and to examine remarkable claims with rigor. On the other hand, science operates on the principle of parsimony—the simplest explanation for all the data is assumed to be the correct one. And we had supported our hypothesis with multiple lines of evidence.

Still, I knew that a single gee-whiz discovery does not have any long-term meaning to science. We had to sequence proteins from other dinosaur finds. When a volunteer accompanying us on a summer expedition found bones from an 80-million-year-old plant-eating duckbill dinosaur called *Brachylophosaurus canadensis*, or "Brachy," we suspected the duckbill might be a good source of ancient proteins even before we got its bones out of the ground. Hoping that it might contain organics, we did everything we could to free it from the surrounding sandstone quickly while minimizing its exposure to the elements. Air pollutants, humidity fluctuations and the like would be very harmful to fragile molecules, and the longer the bone was exposed, the more likely contamination and degradation would occur.

Perhaps because of this extra care—and prompt analyses—both the chemistry and the morphology of this second dinosaur were less altered than Brex's. As we had hoped, we found cells embedded in a matrix of white collagen fibers in the animal's bone. The cells exhibited long, thin, branchlike extensions that are characteristic of osteocytes, which we could trace from the cell body to where they connected to other cells. A few of them even contained what appeared to be internal structures, including possible nuclei.

Furthermore, extracts of the duckbill's bone reacted with antibodies that target collagen and other proteins that bacteria do not manufacture, refuting the suggestion that our soft-tissue structures were merely biofilms. In addition, the protein sequences we obtained from the bone most closely resembled those of modern birds, just as Brex's did. And we sent samples of the duckbill's bone to several different labs for independent testing, all of which confirmed our results. After we reported these findings in *Science* in 2009, I heard no complaints.

Our work does not stop here. There is still so much about ancient soft tissues that we do not understand. Why are these materials preserved when all our models say they should be degraded? How does fossilization really occur? How much can we learn about animals from preserved fragments of molecules? The sequencing work hints that analyses of this material might eventually help to sort out how extinct species are related—once we and others build up bigger libraries of ancient sequences, and sequences from living species, for comparison. As these databases expand, we may be able to compare sequences to see how members of a lineage changed at the molecular level. And by rooting these sequences in time, we might be able to better understand the rate of this evolution. Such insights will help scientists to piece together how dinosaurs and other extinct creatures responded to major environmental changes, how they recovered from catastrophic events, and ultimately what did them in.

MARY H. SCHWEITZER *had already trained to become a high school science teacher when she took a class in paleontology for fun and reignited a childhood interest in dinosaurs. She then earned a Ph.D. in biology from Montana State University in 1995. Today she is an associate professor in the department of marine, earth and atmospheric sciences at North Carolina State University and an associate curator at the North Carolina Museum of Natural Sciences.*

MORE TO EXPLORE

Preservation of Biomolecules in Cancellous Bone of *Tyrannosaurus rex*. Mary H. Schweitzer et al. in *Journal of Vertebrate Paleontology*, Vol. 17, No. 2, pages 349–359; June 1997.

Beta-Keratin Specific Immunological Reactivity in Feather-like Structures of the Cretaceous Alvarezsaurid, *Shuvuuia deserti*. Mary H. Schweitzer et al. in *Journal of Experimental Zoology*, Vol. 285, pages 146–157; August 1999.

Protein Sequences from Mastodon and *Tyrannosaurus rex* Revealed by Mass Spectrometry. John M. Asara et al. in *Science*, Vol. 316, pages 280–285; April 13, 2007.

Dinosaurian Soft Tissues Interpreted as Bacterial Biofilms. Thomas G. Kaye et al. in *PLoS ONE*, Vol. 3, No. 7; July 2008.

Biomolecular Characterization and Protein Sequences of the Campanian Hadrosaur *B. canadensis.* Mary H. Schweitzer et al. in Science, Vol. 324, pages 626–631; May 1, 2009.

Dinosaurs' Living Descendants[*]

By Richard Stone
Smithsonian, December 2010

In a pine forest in rural northeastern China, a rugged shale slope is packed with the remains of extinct creatures from 125 million years ago, when this part of Liaoning province was covered with freshwater lakes. Volcanic eruptions regularly convulsed the area at the time, entombing untold millions of reptiles, fish, snails and insects in ash. I step gingerly among the myriad fossils, pick up a shale slab not much larger than my hand and smack its edge with a rock hammer. A seam splits a russet-colored fish in half, producing mirror impressions of delicate fins and bones as thin as human hairs.

One of China's star paleontologists, Zhou Zhonghe, smiles. "Amazing place, isn't it?" he says.

It was in 1995 that Zhou and colleagues announced the discovery of a fossil from this prehistoric disaster zone that heralded a new age of paleontology. The fossil was a primitive bird the size of a crow that may have been asphyxiated by volcanic fumes as it wheeled above the lakes all those millions of years ago. They named the new species *Confuciusornis*, after the Chinese philosopher.

Until then, only a handful of prehistoric bird fossils had been unearthed anywhere in the world. That's partly because birds, then as now, were far less common than fish and invertebrates, and partly because birds more readily evaded mudslides, tar pits, volcanic eruptions and other geological phenomena that captured animals and preserved traces of them for the ages. Scientists have located only ten intact fossilized skeletons of the earliest known bird, *Archaeopteryx*, which lived at the end of the Jurassic period, about 145 million years ago.

Zhou, who works at the Institute of Vertebrate Paleontology and Paleoanthropology (IVPP) of the Chinese Academy of Sciences in Beijing, believed that the extraordinary bone beds in Liaoning might fill in some of the many blanks in the fossil record of the earliest birds. He couldn't have been more prophetic. In the past 15 years, thousands of exquisitely preserved fossil birds have emerged from the

[*] Article by Richard Stone from *Smithsonian Magazine*, December 2010. Copyright © 2010.

ancient lakebed, called the Yixian Formation. The region has also yielded stunning dinosaur specimens, the likes of which had never been seen before. As a result, China has been the key to solving one of the biggest questions in dinosaur science in the past 150 years: the real relationship between birds and dinosaurs.

The idea that birds—the most diverse group of land vertebrates, with nearly 10,000 living species—descended directly from dinosaurs isn't new. It was raised by the English biologist Thomas Henry Huxley in his 1870 treatise, *Further Evidence of the Affinity between the Dinosaurian Reptiles and Birds*. Huxley, a renowned anatomist perhaps best remembered for his ardent defense of Charles Darwin's theory of evolution, saw little difference between the bone structure of *Compsognathus*, a dinosaur no bigger than a turkey, and *Archaeopteryx*, which was discovered in Germany and described in 1861. When Huxley looked at ostriches and other modern birds, he saw smallish dinosaurs. If a baby chicken's leg bones were enlarged and fossilized, he noted, "there would be nothing in their characters to prevent us from referring them to the *Dinosauria*."

Still, over the decades researchers who doubted the dinosaur-bird link also made good anatomical arguments. They said dinosaurs lack a number of features that are distinctly avian, including wishbones, or fused clavicles; bones riddled with air pockets; flexible wrist joints; and three-toed feet. Moreover, the posited link seemed contrary to what everyone thought they knew: that birds are small, intelligent, speedy, warmblooded sprites, whereas dinosaurs—from the Greek for "fearfully great lizard"—were coldblooded, dull, plodding, reptile-like creatures.

In the late 1960s, a fossilized dinosaur skeleton from Montana began to undermine that assumption. *Deinonychus*, or "terrible claw" after the sickle-shaped talon on each hind foot, stood about 11 feet from head to tail and was a lithe predator. Moreover, its bone structure was similar to that of *Archaeopteryx*. Soon scientists were gathering other intriguing physical evidence, finding that fused clavicles were common in dinosaurs after all. *Deinonychus* and *Velociraptor* bones had air pockets and flexible wrist joints. Dinosaur traits were looking more birdlike all the time. "All those things were yanked out of the definition of being a bird," says paleontologist Matthew Carrano of the Smithsonian National Museum of Natural History.

But there was one important feature that had not been found in dinosaurs, and few experts would feel entirely comfortable asserting that chickadees and triceratops were kin until they had evidence for this missing anatomical link: feathers.

A poor Chinese farmer, Li Yingfang, made one of the greatest fossil finds of all time, in August 1996 in Sihetun village, an hour's drive from the site where I'd prospected for fossil fish. "I was digging holes for planting trees," recalls Li, who now has a full-time job at a dinosaur museum built at that very site. From a hole he unearthed a two-foot-long shale slab. An experienced fossil hunter, Li split the slab and beheld a creature unlike any he had seen. The skeleton had a birdlike skull, a long tail and impressions of what appeared to be feather-like structures.

Because of the feathers, Ji Qiang, then the director of the National Geological Museum, which bought one of Li's slabs, assumed it was a new species of primitive bird. But other Chinese paleontologists were convinced it was a dinosaur.

On a visit to Beijing that October, Philip Currie, a paleontologist now at the University of Alberta, saw the specimen and realized it would turn paleontology on its head. The next month, Currie, a longtime China hand, showed a photograph of it to colleagues at the annual meeting of the Society of Vertebrate Paleontology. The picture stole the show. "It was such an amazing fossil," recalls paleontologist Hans-Dieter Sues of the National Museum of Natural History. "Sensational." Western paleontologists soon made a pilgrimage to Beijing to see the fossil. "They came back dazed," Sues says.

Despite the feathers, the skeleton left no doubt that the new species, named *Sinosauropteryx*, meaning "Chinese lizard wing," was a dinosaur. It lived around 125 million years ago, based on the dating of radioactive elements in the sediments that encased the fossil. Its integumentary filaments—long, thin structures protruding from its scaly skin—convinced most paleontologists that the animal was the first feathered dinosaur ever unearthed. A dozen dinosaurs with filaments or feathers have since been discovered at that site.

By analyzing specimens from China, paleontologists have filled in gaps in the fossil record and traced the evolutionary relationships among various dinosaurs. The fossils finally have confirmed, to all but a few skeptics, that birds descended from dinosaurs and are the living representatives of a dinosaur lineage called the Maniraptorans.

Most dinosaurs were not part of the lineage that gave rise to birds; they occupied other branches of the dinosaur family tree. *Sinosauropteryx*, in fact, was what paleontologists call a non-avian dinosaur, even though it had feathers. This insight has prompted paleontologists to revise their view of other non-avian dinosaurs, such as the notorious meat eater *Velociraptor* and even some members of the tyrannosaur group. They, too, were probably adorned with feathers.

The abundance of feathered fossils has allowed paleontologists to examine a fundamental question: Why did feathers evolve? Today, it's clear that feathers perform many functions: they help birds retain body heat, repel water and attract a mate. And of course they aid flight—but not always, as ostriches and penguins, which have feathers but do not fly, demonstrate. Many feathered dinosaurs did not have wings or were too heavy, relative to the length of their feathered limbs, to fly.

Deciphering how feathers morphed over the ages from spindly fibers to delicate instruments of flight would shed light on the transition of dinosaurs to birds, and how natural selection forged this complex trait. Few scientists know ancient feathers more intimately than IVPP's Xu Xing. He has discovered 40 dinosaur species—more than any other living scientist—from all over China. His office at IVPP, across the street from the Beijing Zoo, is cluttered with fossils and casts.

Xu envisions feather evolution as an incremental process. Feathers in their most primitive form were single filaments, resembling quills, that jutted from reptilian skin. These simple structures go way back; even pterodactyls had filaments of sorts. Xu suggests that feather evolution may have gotten started in a common ancestor of pterodactyls and dinosaurs—nearly 240 million years ago, or some 95 million years before *Archaeopteryx*.

After the emergence of single filaments came multiple filaments joined at the base. Next to appear in the fossil record were paired barbs shooting off a central shaft. Eventually, dense rows of interlocking barbs formed a flat surface: the basic blueprint of the so-called pennaceous feathers of modern birds. All these feather types have been found in fossil impressions of theropods, the dinosaur suborder that includes *Tyrannosaurus rex* as well as birds and other Maniraptorans.

Filaments are found elsewhere in the dinosaur family tree as well, in species far removed from theropods, such as *Psittacosaurus*, a parrot-faced herbivore that arose around 130 million years ago. It had sparse single filaments along its tail. It's not clear why filaments appear in some dinosaur lineages but not in others. "One possibility is that feather-like structures evolved very early in dinosaur history," says Xu, and some groups maintained the structures, while other groups lost them. "But finally in Maniraptorans, feathers stabilized and evolved into modern feathers," he says. Or filaments may have evolved independently at different times. As Sues points out, "It seems that, genetically, it's not a great trick to make a scale into a filament."

Originally, single filaments may well have been for display, the dinosaur equivalent of a peacock's iridescent plumage. Vivid evidence for that theory appeared when scientists unveiled the true colors of 125-million-year-old feathers. Bird feathers and reptile scales contain melanosomes—tiny sacs holding varieties of the pigment melanin. Many paleontologists suspected that dinosaur feathers also contained melanosomes. In Mike Benton's laboratory at the University of Bristol, IVPP's Zhang Fucheng spent more than a year searching for melanosomes in photographs of bird and dinosaur fossils taken with an electron microscope. Zhang's diligence paid off in 2009 when he pinpointed melanosomes in *Confuciusornis* that contained eumelanin, which gives feathers a gray or black tinge, and pheomelanin, which gives them a chestnut to reddish-brown color. The animal's feathers had patches of white, black and orange-brown coloring.

Sinosauropteryx was even more stunning. Zhang found that the filaments running down its back and tail must have made the dinosaur look like an orange-and-white-striped barber pole. Such a vibrant pattern suggests that "feathers first arose as agents for color display," Benton says.

Early feathers could have served other purposes. Hollow filaments may have dissipated heat, much as the frills of some modern lizards do today. Other paleontologists speculate feathers first evolved to retain heat. A telling example comes from fossils of *Oviraptor*—a theropod unearthed in Mongolia that lived around 75 million years ago—squatting over egg-filled nests. *Oviraptors* tucked their legs into the center of the clutch and hugged the periphery with their long forelimbs—a posture bearing an uncanny resemblance to brooding birds keeping their eggs warm. Dinosaurs related to *Oviraptor* were covered with pennaceous feathers, suggesting that *Oviraptor* was as well. "Sitting on a nest like that only made sense if it had feathers" to gently insulate its young, says Sues.

Feathers did, of course, eventually become an instrument of flight. Some paleontologists envision a scenario in which dinosaurs used feathers to help them occupy trees for the first time. "Because dinosaurs had hinged ankles, they could not rotate their feet and they couldn't climb well. Maybe feathers helped them scramble up tree trunks," Carrano says. Baby birds of primarily ground-dwelling species like turkeys use their wings in this way. Feathers may have become increasingly aerodynamic over millions of years, eventually allowing dinosaurs to glide from tree to tree. Individuals able to perform such a feat might have been able to reach new food sources or better escape predators—and pass the trait on to subsequent generations.

One of the most beguiling specimens to emerge from Liaoning's shale beds is *Microraptor*, which Xu discovered in 2003. The bantamweight beast was a foot or two long and tipped the scales at a mere two pounds. *Microraptor*, from the Dromaeosaur family, was not an ancestor of birds, but it was also unlike any previously discovered feathered dinosaur. Xu calls it a "four-winged" dinosaur because it had long, pennaceous feathers on its arms and legs. Because of its fused breastbone and asymmetrical feathers, says Xu, *Microraptor* surely could glide from tree to tree, and it may even have been better at flying under its own power than *Archaeopteryx* was.

Last year, Xu discovered another species of four-winged dinosaur, also at Liaoning. Besides showing that four-winged flight was not a fluke, the new species, *Anchiornis huxleyi*, named in honor of Thomas Henry Huxley, is the earliest known feathered dinosaur. It came from Jurassic lakebed deposits 155 million to 160 million years old. The find eliminated the final objection to the evolutionary link between birds and dinosaurs. For years, skeptics had raised the so-called temporal paradox: there were no feathered dinosaurs older than *Archaeopteryx*, so birds could not have arisen from dinosaurs. Now that argument was blown away: *Anchiornis* is millions of years older than *Archaeopteryx*.

Four-winged dinosaurs were ultimately a dead branch on the tree of life; they disappear from the fossil record around 80 million years ago. Their demise left only one dinosaur lineage capable of flight: birds.

Just when did dinosaurs evolve into birds? Hard to say. "Deep in evolutionary history, it is extremely difficult to draw the line between birds and dinosaurs," says Xu. Aside from minor differences in the shape of neck vertebrae and the relative length of the arms, early birds and their Maniraptoran kin, such as *Velociraptor*, look very much alike.

"If *Archaeopteryx* were discovered today, I don't think you would call it a bird. You would call it a feathered dinosaur," says Carrano. It's still called the first bird, but more for historic reasons than because it is the oldest or best embodiment of birdlike traits.

On the other hand, *Confuciusornis*, which possessed the first beak and earliest pygostyle, or fused tail vertebrae that supported feathers, truly looks like a bird. "It passes the sniff test," Carrano says.

Since the last of the non-avian dinosaurs died out 65 million years ago during the mass extinction that closed the curtain on the Cretaceous period, birds have evolved other characteristics that set them apart from dinosaurs. Modern birds have higher metabolisms than even the most agile *Velociraptor* ever had. Teeth disappeared at some point in birds' evolutionary history. Birds' tails got shorter, their flying skills got better and their brains got bigger than those of dinosaurs. And modern birds, unlike their Maniraptoran ancestors, have a big toe that juts away from the other toes, which allows birds to perch. "You gradually go from the long arms and huge hands of non-avian Maniraptorans to something that looks like the chicken wing you get at KFC," says Sues. Given the extent of these avian adaptations, it's no wonder the evolutionary link between dinosaurs and birds as we know them remained hidden until paleontologists started analyzing the rich fossil record from China.

Chaoyang is a drab Chinese city with dusty streets; in its darker corners it's reminiscent of gritty 19th-century American coal-mining towns. But to fossil collectors, Chaoyang is a paradise, only a one-hour drive from some of the Yixian Formation's most productive beds.

One street is lined with shops selling *yuhuashi*, or fish fossils. Framed fossils embedded in shale, often in mirror-image pairs, can be had for a dollar or two. A popular item is a mosaic in which a few dozen small slabs form a map of China; fossil fish appear to swim toward the capital, Beijing (and no map is complete without a fish representing Taiwan). Merchants sell fossilized insects, crustaceans and plants. Occasionally, despite laws that forbid trade in fossils of scientific value, less scrupulous dealers have been known to sell dinosaur fossils. The most important specimens, Zhou says, "are not discovered by scientists at the city's fossil shops, but at the homes of the dealers or farmers who dug them."

In addition to *Sinosauropteryx*, several other revelatory specimens came to light through amateurs rather than at scientific excavations. The challenge for Zhou and his colleagues is to find hot specimens before they disappear into private collections. Thus Zhou and his colleague Zhang Jiangyong, a specialist on ancient fish at IVPP, have come to Liaoning province to check out any fossils that dealers friendly to their cause have gotten their hands on of late.

Most of the stock in the fossil shops comes from farmers who hack away at fossil beds when they aren't tending their fields. A tiny well-preserved fish specimen can yield its finder the equivalent of 25 cents, enough for a hot meal. A feathered dinosaur can earn several thousand dollars, a year's income or more. Destructive as it is to the fossil beds, this paleo economy has helped rewrite prehistory.

Zhou picks up a slab and peers at it through his wire-rimmed glasses. "Chairman, come here and look," Zhou says to Zhang (who earned his playful nickname as chairman of IVPP's employees union). Zhang examines the specimen and adds it to a pile that will be hauled back to Beijing for study—and, if they are lucky, reveal another hidden branch of the tree of life.

Bibliography

Courtesy of the U.S. Geological Survey
Dinosaur tracks in the Jurassic Moenave Formation, in northern Arizona.

A *Centrosaurus* skull on display at the Museum of Victoria, in Melbourne, Australia. *Centrosaurus* lived in the Late Cretaceous Period, about 75 million years ago.

Books

Brinkman, Paul D. *The Second Jurassic Dinosaur Rush*. Chicago: University of Chicago Press, 2010.

Cadbury, Deborah. *Terrible Lizard: The First Dinosaur Hunters and the Birth of a New Science*. New York: Henry Holt and Co., LLC, 2000.

Chen, Pei-ji and Yuan Wang, eds. *The Jehol Fossils: The Emergence of Feathered Dinosaurs, Beaked Birds and Flowering Plants*. London; Burlington, Mass.: Academic Press, 2008.

Dingus, Lowell and Mark Norell. *Barnum Brown: The Man Who Discovered Tyrannosaurus rex*. Berkeley and Los Angeles, Calif.: University of California Press, Ltd., 2010.

Emling, Shelley. *The Fossil Hunter: Dinosaurs, Evolution, and the Woman Whose Discoveries Changed the World*. New York: Palgrave Macmillan, 2009.

Fastovsky, David E., and David B. Weishampel. *Dinosaurs: A Concise Natural History*. New York: Cambridge University Press, 2009.

———. *The Evolution and Extinction of the Dinosaurs, Second Edition*. New York: Cambridge University Press, 2005.

Fraser, Nicholas. *Dawn of the Dinosaurs: Life in the Triassic*. Bloomington, Ind.: Indiana University Press, 2006.

Holtz, Thomas R., Jr. *Dinosaurs: The Most Complete, Up-to-Date Encyclopedia for Dinosaur Lovers of All Ages*. New York: Random House Children's Books, 2007.

Horner, Jack, and James Gorman. *How to Build a Dinosaur: Extinction Doesn't Have to Be Forever*. New York: Penguin Group (USA) Inc., 2009.

Larson, Peter, and Kristin Donnan. *Rex Appeal: The Amazing Story of Sue, the Dinosaur That Changed Science, the Law, and My Life.* Montpelier, Vt.: Invisible Cities Press, 2002.

Long, John, and Peter Schouten. *Feathered Dinosaurs: The Origin of Birds.* New York: Oxford University Press, 2008.

Lucas, Spencer G., *Dinosaurs: The Textbook.* Boston: McGraw-Hill, 2004.

Maier, Gerhard. *African Dinosaurs Unearthed: The Tendaguru Expeditions.* Bloomington, Ind.: Indiana University Press, 2003.

Manning, Phillip. *Grave Secrets of Dinosaurs: Soft Tissues and Hard Science.* Washington, D.C.: National Geographic Press, 2008.

Naish, Darren. *The Great Dinosaur Discoveries.* Berkeley, CA: University of California Press, 2009.

Novacek, Michael. *Dinosaurs of the Flaming Cliffs.* New York: Anchor Books, 1996.

———. *Time Traveler: In Search of Dinosaurs and Other Fossils from Montana to Mongolia.* New York: Farrar, Straus and Giroux, 2002.

Novas, Fernando E. *The Age of Dinosaurs in South America.* Bloomington, Ind.: Indiana University Press, 2009.

Parker Steve. *Dinosaurus: The Complete Guide to Dinosaurs.* Toronto, Canada: Firefly Books, Ltd., 2003.

Paul, Gregory S. *The Princeton Field Guide to Dinosaurs.* Princeton, N.J.: Princeton University Press, 2010.

Paul, Gregory S., ed. *The Scientific American Book of Dinosaurs.* New York: Saint Martin's Press, 2000.

Sampson, Scott D. *Dinosaur Odyssey: Fossil Threads in the Web of Life.* Berkeley and Los Angeles, Calif.: University of California Press, Ltd., 2009.

Wallace, David Rains. *The Bonehunters' Revenge: Dinosaurs and Fate in the Gilded Age.* New York: Houghton Mifflin, 1999.

Web Sites

Readers seeking additional information on dinosaurs and related subjects may wish to consult the following Web sites, all of which were operational as of this writing.

American Museum of Natural History: Dinosaurs

www.amnh.org/exhibitions/dinosaurs/

From May 14, 2005 to January 8, 2006, New York City's American Museum of Natural History presented, "Dinosaurs: Ancient Fossils, New Discoveries," an exhibition focusing on modern paleontology and the ways in which scientists are using technology to analyze fossils and better understand dinosaurs. Although the exhibition is no longer running, the Web site is still active, and it contains a wealth of information on the biomechanics of theropod and sauropod dinosaurs; the usefulness of studying footprints and trackways; the apparent evolutionary benefits of horns, plates, spikes, and other anatomical features; and various theories regarding what killed the dinosaurs. The site also includes audio and text interviews with seven leading researchers, among them Dr. Mark A. Norell, the museum's chairman and curator-in-charge.

DinoDictionary.com

www.dinodictionary.com/index.asp

Boasting more than 300 entries arranged in alphabetical order, this Web site is an excellent quick reference tool for anyone interested in the height, weight, eating habits, and taxonomical classifications—order, suborder, infraorder—of various types of dinosaurs. A typical entry includes notes on the dinosaur, including where it lived and when it was discovered, as well as a silhouetted image comparing the creature's height to that of modern man. The handy audio component demonstrates how to pronounce words like *Maiasaura*, *Wuerhosaurus*, and *Frenguellisaurus*, lest readers embarrass themselves at their next paleontologist cocktail party.

Discovery Channel: Dinosaurs

dsc.discovery.com/dinosaurs/

Much more than a listing of dinosaur-related Discovery Channel programming—although it is that, too—this Web site contains articles describing the major types of dinosaurs and the prehistoric eras in which they lived. There are also sections devoted to dinosaur news, games, photos, and videos.

LiveScience: Dinosaurs

http://www.livescience.com/topics/dinosaurs/

Founded in 2004, LiveScience aims to be a "trusted and provocative source for highly accessible science, health and technology news for people who are curious about their minds, bodies, and the world around them." The topic page devoted to dinosaurs chronicles new fossil discoveries and scientific theories, keeping dinosaur enthusiasts up to date on the latest research.

National Museum of Natural History: Dinosaurs

paleobiology.si.edu/dinosaurs/

The National Museum of Natural History, a part of the Smithsonian Institution in Washington, D.C., is a treasure trove of dinosaur information and artifacts. While there's nothing quite like visiting the Hall of Dinosaurs and gazing upon the skeletons of *Stegosaurus*, *Allosaurus*, and the other prehistoric creatures featured in the permanent collection, this Web site offers virtual tours—the next best thing. Thanks to "Everything You Want to Know" and "Dinosaur FAQ" sections, the site also provides a helpful introduction to the field of dinosaur study, presenting the latest theories on how dinosaurs lived, evolved, and ultimately died out.

Smithsonian.com: Dinosaur Tracking Blog

blogs.smithsonianmag.com/dinosaur/

With the tagline "where paleontology meets pop culture," this regularly updated blog features postings on everything from new fossil finds to videos of schoolchildren performing interpretative dances about dinosaur extinction. Despite its funny headlines and forays into non-scientific fare, the Web site is as informative as it is entertaining. As of this writing, the blog is maintained by New Jersey State Museum research associate Brian Switek, author of *Written in Stone: Evolution, the Fossil Record, and Our Place in Nature*, published in 2010 by Bellevue Literary Press.

Additional Periodical Articles with Abstracts

More information about dinosaurs can be found in the following articles. Readers interested in additional material may consult the *Readers' Guide to Periodical Literature* and other H.W. Wilson publications.

American Dinosaurs: Who and What Was First? Keith Stewart Thomson. *American Scientist* v. 94 pp209–11 May/June 2006.

The identity of the person who discovered the first North American dinosaur is debatable, Thomson writes. Joseph Leidy of Philadelphia is credited with discovering the first North American dinosaur remains in Montana in 1856, however, in 1836, Edward Hitchcock, president of Amherst College in Massachusetts, described dinosaur trackways in Connecticut. The main argument against recognizing Hitchcock as the first discoverer, Thomson remarks, is that his finds were impressions made by dinosaurs rather than bony bodily remains. Moreover, it is reasonably certain that the first discovery of a dinosaur thigh bone was made in 1787 by Timothy Matlack, a Philadelphia merchant, and Caspar Wistar, a physician and anatomist, Thomson adds.

Dinosaurs as a Cultural Phenomenon. Keith Stewart Thomson. *American Scientist* v. 93 pp212–14 May/June 2005.

The modern popularity of dinosaurs is due to the creatures themselves and to astute showmanship and media savvy, Thomson contends. From the beginning, the author says, some dinosaur sleuths promoted their discoveries in ways that other fossil hunters did not. In 1897 and 1898, American newspapers presented images of *Brontosaurus* in a theatrical pose against a backdrop of skyscrapers, and the reception given to these images firmly illustrated the potential of dinosaurs to capture public interest. Although dinosaurs are surrounded by extraordinary marketing and publicity, Thomson observes, such efforts would have fizzled out by now unless dinosaurs were fundamentally interesting. Dinosaurs' hold on the collective imagination of the public may be partly because dinosaurs are the most paradoxical of all extinct organisms, the author suggests. Paleontology, and dinosaur paleontology in particular, is the most accessible aspect of the concept of evolution, Thomson adds, raising questions about what will happen if dinosaurs ever lose their appeal.

Arrested Development: Using Technology to See Ancient Embryos. Myrna E. Watanabe. *BioScience* v. 60 pp490–94 July/August 2010.

Advances in microscopy and other imaging technologies are enabling scientists to study the contents of fossilized eggs, Watanabe reports. Creative vertebrate paleontologists and geologists are adapting methods employed in oil geology, invertebrate paleontology, and biomedical research to study the characteristics of dinosaur eggshells. In some cases, they are able to produce images of the embryonic material found inside shells.

Dino Was Prehistoric Hunk. *Current Science*. v. 96 p15 September 17, 2010.

In this article, the writer describes the new dinosaur, *Medusaceratops lokii*, recently discovered by fossil hunters in Montana. The dinosaur lived about 78 million years ago, measured 6 meters (20 feet) long, and weighed about 20 tons. Its snout ended in a beak like that of a parrot and a curved spike as long as a baseball bat protruded over each eye. In addition, rising from its neck was an enormous frill, or collar, rimmed with snake-like hooks.

Gerta Keller: 'Volcanoes Killed the Dinosaurs.' Lindsay Patterson. *EarthSky* (on-line) 2009.

Although scientists have long believed a meteorite impact in the Yucatan caused a mass extinction of species, including the dinosaurs, Patterson writes, geologist Gerta Keller of Princeton disagrees. Keller's research suggests the Yucatan impact occurred 300,000 years after the dinosaurs disappeared, and she believes volcanoes are to blame for the extinction. Eruptions on India's Deccan Plateau between 63 and 67 million years ago released huge amounts of sulfur dioxide into the air, Patterson observes. Moreover, by studying geologic core samples from the area, Keller's team discovered less evidence of life with each subsequent volcanic flow, lending credence to the theory.

Steve Brusatte Says Dinosaurs Were Not Special, Just Lucky. Lindsay Patterson. *EarthSky* (on-line) 2009.

In this article, Patterson asks Steve Brusatte, a Ph.D. student at Columbia University and the American Museum of Natural History in New York, why scientists consider dinosaurs "lucky." His response: They survived a mass extinction event at the end of the Triassic period, a time when they were in direct competition with crurotarsans, or ancient crocodiles. According to Brusatte, crurotarsans were more diverse and more adapted, and yet for some reason, the dinosaurs survived and the crurotarsans didn't. While it stands to reason dinosaurs must have had some feature that helped them survive, Brusatte's research suggests they were not innately superior. The lesson learned, he concludes, is that evolution is not orderly or predictable, and that luck plays a hand as well.

Goseong: The Home of Dinosaurs. Hyungyoon Kim. *Koreana* v. 23 pp68–75 Summer 2009.

As Kim explains, there are numerous signs of the presence of dinosaurs along the valleys of Goseong, South Korea. More than 4,300 sets of dinosaur footprints have been uncovered in the Goseong area, which has been listed as one of the three premiere sites in the world for dinosaur footprint fossils. Scientists believe a natural disaster resulted in the area's ground surface becoming suddenly submerged, and that the footprints of dinosaurs were preserved under water.

Predator or Putz? Kate Lunau. *Maclean's* v. 122 pp58–9 November 16, 2009.

With its bone-crushing teeth and gigantic proportions, Lunau writes, no creature is more fearsome in the public imagination than the mighty *Tyrannosaurus rex*. Sixty-five million years after it became extinct, however, *T. Rex* is having an identity crisis: In recent months, much of what is known about this iconic monster has been "flipped on its ear," asserts Stephen Brusatte, a vertebrate paleontologist at the American Museum of Natural History in New York. Indeed, the *T. Rex's* size, speed, and eating habits have all been thrown into question, Lunau explains, leaving dinosaur fans wondering if the *T. Rex* really was so fearsome after all.

The Quest to Build a Dinosaur. Kate Lunau. *Maclean's* v. 122 pp40–43 August 24, 2009.

Scientists believe they are close to bringing dinosaurs back to life, Lunau reports. By making a few genetic tweaks to its modern-day ancestor, the bird, world-famous paleontologist Jack Horner wants to hatch a dinosaur straight from a chicken egg. With Horner's encouragement, McGill University paleontologist Hans Larsson is experimenting with chicken embryos to create the creature Horner describes as a "chickenosaurus."

The Real Jurassic Park. Peter Gwain. *National Geographic* v. 214 pp104–15 July 2008.

Working in northwestern China's Junggar Basin, paleontologists James Clark and Xu Xing, supported by the National Geographic Society, have discovered a remarkable collection of fossils, Gwain notes. Their excavations have provided new insight into an obscure period in Earth's geologic history, a brutal era that lasted from approximately 165 million to 155 million years ago and saw the continents breaking up and dinosaurs experiencing a burst of evolution. The division of the landmasses meant animals became isolated from each other, the author writes, which led to many new branches sprouting on the dinosaur family tree. Nevertheless, scientists have been confounded by a lack of terrestrial fossils from the period, making the findings of Clark and Xu all the more vital.

How Did Dinosaurs Begin To Fly? From the Ground Up. Luis M. Chiappe. *Natural History* v. 114 pp54–55 May 2005.

According to Chiappe, avian flight probably originated in theropod dinosaurs that took off from the ground. A primitive bird or theropod could have increased its thrust, or force in the direction of its run, by flapping its feathered forelimbs, the author writes. The increased thrust would have boosted the running speed of the animal and simultaneously increased its lift, the upward force produced by the air moving across the animal's forelimbs. With the increased lift, the force between the bird's hind limbs and the ground would have declined until it reached zero and the creature achieved flight, Chiappe states.

Dinosaurs Among Us? Paul Chambers. *New Scientist* v. 186 p46 May 21, 2005.

This article is part of a special section on dinosaur discoveries. Although it is clear that birds evolved from dinosaurs, Chambers writes, much remains to be learned about bird evolution. One mystery is why modern birds survived while a group called opposite birds, which contained most Mesozoic-era birds, disappeared with the dinosaurs. There is also conflicting evidence, Chambers explains, regarding when modern birds first appeared, with many paleontologists believing that they evolved around 55 million years ago, while others date the origins as far back as 100 million years ago.

A Big Bone Scam. Jerry Adler. *Newsweek*. v. 153 p56 March 23, 2009.

Dinosaur specialists are saddened but not shocked by the theft of a fossil from a ranch in north central Montana, Adler reports. Nate Murphy, a self-instructed, freelance fossil hunter who managed a small research station and museum in the nearby town of Malta, Montana, was a true specialist, the finder in 2000 of one of the best-preserved large dinosaur fossils ever uncovered, a duck-billed *Brachylopho-saurus* he nicknamed Leonardo. In 2006, according to court documents, however, Murphy quietly unearthed and spirited away another amazing fossil from the same ranch—a well-preserved skeleton of a *Bambiraptor*, a fox-scale, bird-like dinosaur, first identified in 1995 and still extremely rare—then lied about where he had found it and had casts made of the specimen, which he sold for a profit.

The Hunchback Of Central Spain. Gwyneth Dickey. *Science News* v. 178 p16 October 9, 2010.

A recently discovered dinosaur species has hints of feather-like appendages on its arms and a strikingly unusual hump on its back, Dickey states. The dinosaur, *Con-cavenator corcovatus*, was found exquisitely preserved in dense limestone at a fossil site in Cuenca, Spain, in 2003. Vertebrate paleontologist Francisco Ortega of the Universidad Nacional de Educación a Distancia in Madrid and his colleagues published the first description of the dinosaur in a issue of *Nature* after seven years of painstaking removal. The feather-like appendages suggest that feathers evolved in more primitive dinosaurs than formerly thought.

Dinosaur Dads as Caretakers. Laura Sanders. *Science News* v. 175 p14 January 17, 2009.

In an issue of *Science*, Varricchio and colleagues present their findings on the pa-rental division of labor in dinosaurs, Sanders notes. Males of modern avian groups that are closely related to dinosaurs, such as emus and ostriches, shoulder the bulk of childcare responsibilities, Sanders writes. Moreover, she adds, parents share childcare duties in the great majority of modern bird species. Varricchio and col-leagues analyzed fossils of an adult Citipati dinosaur, fossilized in the brooding po-sition on a clutch of eggs, and an adult Troodon dinosaur, fossilized while at a nest site. The researchers found no female-specific markers in either fossil, leading them to conclude that the two fossils were likely males, Sanders explains. Furthermore, she adds, the unusually large size of the dinosaur clutches—from 22 to 30 large eggs per clutch—was similar to the clutches of birds that exhibit paternal care.

Dino Demise Traces to Asteroid-Family Breakup. R. Cowen. *Science News* v. 172 p148 September 8, 2007.

Cowen states that the impact responsible for the dinosaurs' extinction was the result of a collision between two asteroids, according to an article by Bottke and colleagues published in *Nature*. The researchers studied a group of asteroids, the so-called Baptistina family, Cowen reports, at a location in the asteroid belt between Mars and Jupiter where stray objects could easily be projected toward Earth. Analysis of the Baptistina asteroids' paths suggests that they were the products of a collision 160 million years ago. Some collision debris would have reached Earth, Cowen writes, doubling the number of objects striking the planet. The composition of the crater from the Chicxulub impact in Mexico, responsible for the extinction of the dinosaurs 65 million years ago, suggests that the object involved originated from the Baptistina family, Cowen adds.

Dinos Burrowed, Built Dens. Sid Perkins. *Science News* v. 172 p259 October 27, 2007.

Paleontologists have found the first indisputable evidence that some dinosaurs maintained an underground lifestyle for at least part of their lives, Perkins reports. Paleontologist David J. Varricchio of Montana State University in Bozeman and his colleagues found the remains of an adult and two juvenile dinosaurs of a completely new species—dubbed *Oryctodromeus cubicularis*—in an unusual sandstone mass protruding from the surrounding rock in southwestern Montana. Varricchio and his colleagues suggest that the anomalous mass of sandstone represents a sudden influx of material that filled in a burrow, trapping its occupants. The researchers also propose that this newly discovered combination of burrow and bodily remains constitutes the first strong evidence of a digging, denning dinosaur.

Big and Birdlike. Sid Perkins. *Science News* v. 171 p371 June 16, 2007.

Paleontologists have discovered the remains of a huge, fast-growing *Oviraptor* dinosaur with body proportions unlike those predicted by the evolutionary trends that characterize its smaller relations, Perkins states. In the journal *Nature*, Xu et al. report that the bird-like creature, which has been given the genus name *Gigantoraptor*, stood 3.5 meters tall at the shoulder, was about 8 meters long, and probably weighed about 1.4 metric tons. Other Oviraptor dinosaurs weighed no more than 40 kilograms, Perkins writes, and the researchers originally thought that the remains were from a sauropod or a tyrannosaur.

T. Rex Fossil Yields Recognizable Protein. Sid Perkins. *Science News* v. 171 p228 April 14, 2007.

A new analysis of a 68-million-year-old *Tyrannosaurus rex* leg bone has revealed substantial protein remnants, Perkins writes. In *Science*, paleontologist Mary Schweitzer and colleagues report that a number of lines of evidence point to the presence of the protein collagen, which is the major non-mineral component of living bone. The amino acid sequences of the *T. rex* collagen more closely matched those found in chickens than in other animals that were tested, Perkins notes, adding to the evidence that modern birds are closely related to dinosaurs.

T. Rex Vision Was Among Nature's Best. Eric Jaffe. *Science News* v. 170 p3–4 July 1, 2006.

Tyrannosaurus rex had some of the best vision in animal history, according to a report in the summer *Journal of Vertebrate Paleontology*. As Jaffe explains, Kent Stevens used facial models of seven types of dinosaurs to reconstruct the binocular range of *T. rex*. He found that the dinosaur had a binocular range of 55°, which is wider than that of modern hawks. *T. rex* also evolved features that enhanced its vision: its snout grew lower and narrower, cheek grooves cleared its sight lines, and its eyeballs enlarged. Stevens determined that T. rex might have had visual acuity as much as 13 times that of humans and had a limiting far point of 6 km, in comparison with a human's far point of 1.6 km.

15-Horned Dinosaur Unearthed. *Science Scope*. v. 34 p11 November 2010.

According to the author, two new species of horned dinosaurs were discovered recently in Grand Staircase–Escalante National Monument, located in southern Utah. The massive herbivores—*Utahceratops getyi* and *Kosmoceratops richardsoni*—were inhabitants of the "lost continent" of Laramidia, formed when a shallow sea flooded the central region of North America in the Late Cretaceous period.

The Strange Lives of Polar Dinosaurs. Mitch Leslie. *Smithsonian* v. 38 pp68–74 December 2007.

Paleontologist Tom Rich and other scientists working in Australia, Alaska, and Antarctica have discovered the remains of dinosaurs that lived in environments that were cold for at least part of the year, Leslie reports. Known as polar dinosaurs, the creatures also had to endure long periods of darkness, up to six months each winter. The evidence that dinosaurs survived in cold conditions, and perhaps scrunched through snow and slid on ice, is at odds with what scientists know about how the animals lived, Leslie explains. Rich, of Melbourne's Museum Victoria, was not the first person to unearth polar dinosaurs, but he and a handful of fellow paleontologists are providing more information about the animals' lives and what their environments were like.

Taking a Dinosaur's Temperature. Mitch Leslie. *Smithsonian* v. 38 p74 December 2007.

Polar dinosaurs have reignited one of paleontology's great debates: whether the dinosaurs were cold-blooded or warm-blooded, Leslie writes. David Fastovsky of the University of Rhode Island argues that polar species support the theory that dinosaur metabolism differed from that of modern reptiles, and he points out that reptiles are not found in frigid climates today. The evidence hints at warm-bloodedness for some dinosaurs, according to Museum Victoria's Tom Rich, who contributed to the debate by sending pieces of two important Australian specimens to the South African Museum in Cape Town, where Anusuya Chinsamy-Turan tested them for lines of arrested growth, or LAGS, which, like tree rings, indicate that growth ceased temporarily. Modern reptiles that live in seasonal environments display LAGS, as do mammals that hibernate, whereas birds and other mammals typically do not, Leslie notes.

Beyond T. Rex. Charles W. Petit. *U.S. News & World Report* v. 133 pp42–46 July 1, 2002.

Dinosaur discovery has been moving at such a pace that it is dissolving the mental niche, half fact and half metaphor, where most people have parked the extinct animals, Petit contends. A group of bigger, more powerful meat-eaters is threatening *Tyrannosaurus rex*'s perch, and *Brontosaurus*, properly known as *Apatosaurus*, has been overshadowed by plant-eating giants that may have weighed five times as much, according to Petit. The re-created Mesozoic landscape now features entirely new kinds of creatures. The remarkable reordering of dinosaur hierarchy is due in large part to geography, Petit writes, explaining that almost all of the celebrity dinosaurs lived and died in what became the American West, one of the first areas paleontologists made a serious, fruitful, and well-publicized search. In recent years, however, fossil hunters have taken the search to other continents, and with major discoveries in China, Africa, and South America, they have discovered a new world of dinosaurs.

Index

About the Editor

JOHN J. MEIER is a science librarian at the Physical and Mathematical Science Library at the Penn State University Libraries. He earned both bachelor's and master's degrees in electrical and computer engineering from the Carnegie Institute of Technology at Carnegie Mellon University. He received his master's of library and information science (MLIS) from the School of Information Sciences at the University of Pittsburgh. Before coming to Penn State, he worked at the University of New Orleans' Earl K. Long Library and at the University of Pittsburgh at Bradford's Hanley Library. He lives in State College, Pennsylvania, with his wife and two children.